Pelicans
to Polar Bears

Pelicans to Polar Bears
Watching Wildlife in Manitoba

Catherine Senecal

Heartland

Heartland Publications
Winnipeg, Canada

Printed in Manitoba, Canada

Hecla/Grindstone Provincial Park, opposite

Copyright ©1999 by the Province of Manitoba

All rights reserved. The use of any part of this publication reproduced, transmitted in any form or by any means – electronic, mechanical, photocopying, recording, or otherwise – or stored in a retrieval system without the prior written consent of the Province of Manitoba and the publisher, Heartland Associates Inc., (or in the case of photocopying, a licence from the Canadian Reprography Collective, or CANCOPY) is an infringement of the copyright law.

Canadian Cataloguing in Publication Data

Senecal, Catherine M.

Pelicans to polar bears: watching wildlife in Manitoba

ISBN 1-896150-02-0

1. Wildlife viewing sites – Manitoba – Guidebooks.
2. Wildlife watching – Manitoba – Guidebooks. 3. Manitoba – Guidebooks. I. Title.

| QL221.M3 S45 1999 | 591.97127 | C99-920014-3 |

Pelicans to Polar Bears was created with funding support provided by the Department of Industry, Trade and Tourism and with the assistance of the Manitoba Wildlife Viewing Guide Steering Committee.

Heartland Publications Inc.
PO Box 103, RPO Corydon
Winnipeg, Manitoba, Canada R3M 3S7

10 9 8 7 6 5 4 3 2 1

Credits

Contributing authors
Dolores Haggarty and Doug Whiteway

Editor
Barbara Huck

Research
Rudolf Koes

Research assistance
Karen Paquin and Erin Huck

Graphic design
Michael Fedun

Design layout, cover design and maps
Dawn Huck

Manitoba Wildlife Viewing Guidebook Steering Committee
Jan Collins (Chair), Manitoba Industry, Trade and Tourism
Glen Suggett (Vice-chair), Wildlife Branch, Manitoba Natural Resources
Donna Danyluk, Manitoba Naturalists Society
Darlene Garnham, Manitoba Wildlife Federation
Jean Horton, Brandon Natural History Society
Rick Hurst, Parks and Natural Areas Branch, Manitoba Natural Resources
Linda Sutterlin-Duguid, Canadian Heritage, Parks Canada
Charlie Taylor, Manitoba Lodges and Outfitters Association
Various representatives of Manitoba Information Resources Division

Prepress
Embassy Graphics, Winnipeg, Canada

Printing
Printcrafters, Winnipeg, Canada

Cover photograph
American white pelican, by Guy Fontaine

Back cover photographs
The Lily Pond, Whiteshell Provincial Park, by Ian Ward
Yellow-headed blackbird and polar bear, both by Dennis Fast

Acknowledgements

With the collaboration that comes with such a project, the lines of responsibility often blur. Many people were involved in this book, including researchers, wildlife experts, editors, photographers, illustrators and designers. Key to the success of the project were the members of the Manitoba Wildlife Viewing Guide Steering Committee and particularly its knowledgeable and dedicated volunteers.

Donna Danyluk of the Manitoba Naturalists Society, Jean Horton of the Brandon Natural History Society and Charlie Taylor of the Manitoba Lodges and Outfitters Association contributed insights, expertise, enthusiasm and an eye for detail over many months as the book went from conception to completion. They attended dozens of meetings, poured over manuscripts and gave thoughtful advice on everything from birds, flowers and photographs to titles and topography. Heartland and I are indebted to them.

I am also grateful for the assistance of a large number of subject experts and knowledgeable amateur naturalists who were of great assistance throughout the process. They include Rick Andrews, Larry Bidlake, Allan Bindle, Nancy Bremner, Rob Bruce, Alice Chambers, Doug Clark, Michael Cobus, Gene Collins, Pat Cormie, Vince Crichton, Heather Davies, Richard Davis, Ken DeSmet, Jackie Dixon, Jack Dubois, Jim Duncan, Cam Elliott, Elaine Gauer, Larry Geisbrecht, Roger Glufka, Gloria Goulet, Jason Greenall, Sylvia Grier, Helios Hernandez, George Holland, Lanette Huculak, Laurence Huculak, Cam Hurst, Bob Jones, Ken Kingdon, Bill Koonz, Donna Kurt, Ron Larche, Robbin Lindsay, Rod MacCharles, Jillian Maguet, Kurt Mazur, Janet Moore, Tom Moran, Lynn Nolden, Dennis Pankiw, Ken Porteous, Bill Preston, Liz Punter, Laura Reeves, Deidre Reis, Dave Roberts, Eric Saunders, Irwin Schellenberg, Ken Schykulski, Rosalie Sigurdson, Hugh Skinner, Peter Taylor, Barry Verbiwski, Ian Ward, Daniel Weedon, Kent Whaley and Doug Yurik.

Many books proved helpful during my research. Some are listed in the Suggested Reading section on page 253. Others that were useful include: *A Field Guide to the Birds of Eastern and Central North America*, by Roger Tory Peterson; *The Birds of Canada* by W. Earl Godfrey; *Manitoba, An Adventure in Nature* by Barbara Huck and Doug Whiteway; *A Birder's Guide to Churchill* by Bonnie Chartier; *Birder's Guide to Southwestern Manitoba* by Calvin W. Cuthbert et al.; *Birder's Guide to Southeastern Manitoba*, by Norman J. Cleveland et al.; *Watching Birds* by Roger F. Pasquier; *The Birds of Manitoba* by Ernest Thompson Seton, and *Watching Wildlife* by Mark Damian Duda.

Arctic hare, opposite

Final thanks are for my husband, Jim Arthur, who came up with the general title concept and supported me along the way. Also, I would like to recognize our three-year-old son, Jasper, who by saying things like "What's the duck next to the redhead?", shows promising signs of becoming a wildlife watcher.

Catherine Senecal
February 1999

Contents

Watching Wildlife in Manitoba — 10

Winnipeg Region – Wildlife in River City — 15

Trip Planning	17	La Barrière Park	34
Assiniboine Park & Forest	19	Birds Hill Provincial Park	36
Fort Whyte Centre	22	Bur Oak Trail	39
Living Prairie Museum	26	Cedar Bog Trail	40
Beaudry Provincial Park	28	White-tailed Deer Trail	43
Oak Hammock Marsh WMA	30		

Eastern Manitoba – Shield & Forest — 45

Trip Planning	47	White Pine Trail	68
Grand Beach Provincial Park	49	Pine Point Trail	69
Ancient Beach Trail	51	Mantario Hiking Trail	70
Dunes & Beaches	53	Seven Sisters Falls	72
Wild Wings Trail	54	Pinawa	74
Spur Woods WMA	56	Nopiming Provincial Park	77
Buffalo Point,		Fire of 'Eighty-Three Trail	80
Birch Point & Moose Lake	58	Walking on Ancient Mountains Trail	82
Whiteshell Provincial Park	61	Bird River Canoe Route	84
Alf Hole Goose Sanctuary	65	Manigotagan Canoe Route	85
Lily Pond Trail	66	Atikaki Provincial Wilderness Park	87
McGillivray Falls Trail	67		

Central Manitoba – The Legacy of Lake Agassiz — 91

Trip Planning	93	Mantagao Lake WMA	106
Netley-Libau Marsh	94	Long Point & Katimik Lake	108
Narcisse WMA	96	Delta Marsh	110
Hecla/Grindstone Provincial Park	99	West Lake Sites	114
Grassy Narrows Marsh	101	Portia Marsh & Bluff Creek	116
Lighthouse Trail	103	Long Island Bay &	
Shoal Lakes	104	Lake Winnipegosis Islands	117
Dog Lake WMA &		Waterhen Wood Bison	118
Lake Manitoba Narrows	105		

Dennis Fast

Southern Manitoba – The Grasslands — 121

Trip Planning	123	Brandon Hills WMA	152
Tall Grass Prairie Preserve	125	Turtle Mountain Provincial Park	154
Crescent Lake	128	Turtle's Back Trail	159
Pinkerton Lakes	131	Souris River Bend WMA	160
Spruce Woods Provincial Park	132	Whitewater Lake WMA	163
Isputinaw Trail	138	Alexander-Griswold Marsh	165
Spirit Sands – Devil's Punch Bowl	140	Oak Lake-Plum Lakes	166
Epinette Creek Trails	143	Poverty Plains	168
Assiniboine River Canoe Route	145	Broomhill WMA	170
Pembina Valley WMA	147	Mixed Grass Prairie Preserve	171
Holmfield WMA	149	Pierson WMA	173
Assiniboine River Corridor - Brandon	150		

Western Manitoba – Prairie Mountains & Parklands — 175

Trip Planning	177	Asessippi Provincial Park	195
Minnedosa Potholes		Ancient Valley Trail	197
& Proven Lake WMA	178	Frank Skinner Arboretum	198
Big Valley	180	Duck Mountain Provincial Park	201
Riding Mountain National Park	183	Baldy Mountain	204
Wasagaming Area	187	Duck Mountain Forest Centre	206
Highway 19	188	Wapiti Trail	207
Lake Audy Plains Bison	189	Shining Stone Trail	208
Highway 10	190	Shell River Valley Trail	210
Birdtail Bench	191	Porcupine Provincial Forest	212
Vermillion Park	192	Overflowing River	213
		Tom Lamb & Saskeram WMAs	214

Northern Manitoba – The Boreal Forest — 217

Trip Planning	219	Pisew Falls Provincial Park	227
Athapapuskow Lake	220	Seal River	229
Grass River Provincial Park	223	Northern Caribou Country	232
Karst Spring Trail	225		

Coastal Manitoba – The Hudson Bay Lowlands — 235

Trip Planning	237	York Factory,	
Churchill River Estuary	238	the Nelson & Hayes Rivers	248
Akudlik Marsh	242	Cape Tatnam WMA	250
Wapusk National Park			
& Cape Churchill WMA	245		

Designation of Lands in Manitoba — 252
Suggested Reading — 253
Some Sought-after Species & Where to Find Them — 254
Manitoba's Top Wildlife Viewing Sites — 256

Plains bison, opposite

Watching Wildlife in Manitoba

From the rugged Canadian Shield of the east – where forests abound with black bears and warblers – to the lush prairie mountains of the west, from the starkly beautiful landscape of the Hudson Bay coastline – home to beluga whales and caribou – to the grasslands of the south, Manitoba offers some of the most varied and accessible wildlife viewing in Canada. This book features 100 viewing sites, some well-known, some less so. Whether you choose to drive the province's highways, binoculars at hand, or fly over forest and marshland to reach secluded lakes, whether you paddle its scores of rivers or hike trails chosen from hundreds of options, you will find sites in this book to suit your interests.

Snowy owl

Spring brings more than a million birds through the province – hundreds of thousands of Canada and snow geese, an amazing array of warblers and remarkable congregations of hawks. In summer you may find a solitary wolf ambling across a highway or a ringed seal bobbing its head up in a northern river estuary. Fall is prime time to see and hear elk or see Churchill's world famous polar bears. Winter offers the chance to ski past a bull moose or see a snowy owl fly silently over the prairie. These are only a few of the wildlife experiences described within.

Using this guide

Sites in this book are numbered according to the page on which they appear. These numbers can be found both in the Contents on page 8, in the alphabetical list of sites on page 256 and on the maps that precede each region. If you are looking for the most sought-after wildlife species and want to know the best places to view them, refer to the list on page 254. If you plan to travel in a certain region, check the Trip Planning page for that area for suggestions on one-day to one-week or longer excursions or driving loops.

The art of wildlife viewing

Manitoba is rich with wildlife viewing opportunities, yet there are no guarantees when watching wildlife. Animals tend to avoid people and veteran wildlife watchers stress that success takes patience and determination. Understanding the habits of the animals you seek is a great beginning, but the best rewards come from looking in the right place at the right time in the right way.

The right place

Within its range, an animal lives in a habitat that can support it. Moose, for example, love well-forested areas with plenty of marshes. During breeding season, pelicans look for lakes with islands. Good field guides will describe the typical habitat for each species. The regular use of such guides assists wildlife viewers in recognizing the important link between habitats and species, and ultimately in knowing what animals to expect in each place.

The right time

Some animals are active during the day, others at night. Some can be seen only at certain times of the year, others year-round. Typically, dawn and dusk are good viewing times for most birds and large mammals. You may also find increased activity before or after a rainfall.

For birds, timing is crucial to see peak numbers at the height of migration. For example, hundreds of thousands of Canada geese and other waterfowl come through Manitoba in peak numbers between mid-April and mid-May or between late August and early October. Spring is the best time for a beginner to learn bird identification as birds call frequently and display their courtship behaviors and distinctive, often brilliant, breeding plumages.

In this guide, best times to see particular species are indicated.

The right way

Learn to recognize the calls, markings, behaviors and habitats for each animal. Many birds look and sound very different during the breeding season. The behavior of elk, moose and other ungulates changes dramatically during the fall mating season. Males can be aggressive. In spring, females with young can also be dangerous.

If you don't see wildlife, look for their tracks and other signs instead. Narrow trails indicate routes to food, shelter or water. Trees may be marked with antler rubbings or deep scratches from bear claws. Animals leave tracks in snow, wet sand or soft ground. Scat, or animal excrement, can identify a particular species as clearly as a calling card. Also, watch for nests or beds in the grass, or fur or hair caught on trees or bushes.

For optimum viewing of both birds and mammals, move slowly and quietly. Avoid sudden movements. Watch for horizontal shapes in a vertical forest and don't forget to look up and down as well as ahead. Be alert for movements in grasses, bushes and trees. Use good binoculars or a spotting scope. Experienced photographers use telephoto lenses, tripods and, often, blinds. Except when dealing with endangered species, try imitating animal calls. Sign up for an excursion with a professional guide, park interpreter or experienced naturalist.

Wild ginger

Linda Fairfield

Responsible Wildlife Viewing

Respecting wildlife and habitats

Who's watching whom? An Arctic tern comes in for a closer look.

Driving the posted speed limit and using caution in signed areas will help to avoid vehicle collisions with wildlife. When planning hiking trips, take time to learn how to behave around large mammals, such as bears, elk, bison or moose. Never feed or touch a wild animal of any age, even if it appears hurt or abandoned. Contact the nearest district office of Manitoba Natural Resources, or the warden service if in a national park. Leave nests, nesting birds, denning mammals and young of any species alone. A young animal's mother, likely nearby, may see your presence as a threat to her young. Consider leaving pets at home because they may threaten or aggravate wildlife. Keep children in sight and never encourage them to pet, feed or pose with wildlife.

Never approach a wild animal to within its fight-or-flight zone, the distance from humans at which an animal feels threatened and shows some type of stress. Birds may bob their heads, ungulates may paw the ground, bears may snort or woof. Depending on the species and whether it is protected or hunted in its habitat, this zone can range from one to hundreds of metres. As a general rule, stay at least 30 metres away from large ungulates and 100 metres away from bears.

Contact provincial park staff, the Manitoba Wildlife Branch or Parks Canada for more information about avoiding problems with bears.

Respect the habitat you are in by carrying litter out and using washrooms or pit toilets where provided. Do not remove any plants, trees, rocks or artifacts.

Respecting the rights of others

Always get permission before entering private land. The public has the right to access public land in the province, and Crown land maps showing public lands are available by calling the Land Information Centre of Manitoba Natural Resources (Map Sales).

While exploring Manitoba, you may encounter hunters and their families also enjoying the outdoors. Our province's hunters take pride in hunting ethically, safely and responsibly. Hunting is a time-honored tradition with deep historical roots providing a wide range of social, economic and cultural values. Hunters are proud of their history of conservation achievements and the major role they continue to play with respect to abundant, healthy wildlife populations.

Other Cautions

Mosquitoes: in late spring and summer, mosquitoes may be found everywhere in Manitoba. Wear gloves and netted head coverings or bug shirts. You could take along a deet-based repellent (a repellent using diethyl toluamide or d.t.). Experienced wildlife watchers often seek a windy point, which mosquitoes avoid, to set up camp.

Blackflies: the good news about blackflies is that they seem to like open spaces – which means they won't invade your tent – and they disappear at nightfall. Blackflies may be found in the province wherever there is clean, flowing water, since this is where the larvae grow and develop.

Ticks: the American dog tick, or wood tick, which is about half the size of a small fingernail, is prevalent in grassy and forested areas from southern Manitoba north into eastern, central and western Manitoba from May to early July. To help avoid ticks, wear light-colored clothing and tuck pants into socks when hiking through dense brush, overgrown trails or grassy areas. Deet is effective against ticks. A tick collar may help your pet. When you return home, check your entire body including your scalp for ticks. If you find an attached tick, use either a pair of blunt tweezers or your fingers (covered with a tissue or latex gloves), grasp the tick as close to the skin as possible and pull straight out with gentle, even pressure. Destroy the tick in alcohol and cleanse the bite area with alcohol or hydrogen peroxide.

What to Take

In addition to this guide, wildlife watchers should be armed with Manitoba maps (highway, topographical and/or air photos depending on your destination), field guides (see Suggested Reading on page 253), a detailed map for canoe routes or unmarked hiking trails and a compass or GPS device. Maps are available from the Land Information Centre (Map Sales outlet) in Winnipeg. Provincial park maps are available at Manitoba Natural Resources offices. Binoculars or spotting scopes are a great help. In summer, wear hats and light-weight clothing and bring water and sunscreen. In winter, wear lined boots, layers of clothing and effective head, face and hand coverings.

For more information about wildlife, travel services, accommodation and more, call **Travel Manitoba** at **1-800-665-0040 ext. EW9** or visit **www.travelmanitoba.com**. For provincial parks and natural areas, visit **www.gov.mb.ca/natres**. For national parks in Manitoba, call **Parks Canada** at **1-888-748-2928** or visit **www.parkscanada.pch.gc.ca**. For information on birding, natural history outings and programs visit the **Manitoba Naturalists Society** at **www.wilds.mb.ca/mns** or telephone **(204) 943-9029**. For Manitoba lodges and outfitters visit **www.mloa.com**. Canoe route maps, topographical maps and air photos are available from the **Land Information Centre, Manitoba Natural Resources** at **(204) 945-6666**.

The blacklegged or deer tick, is smaller than the common wood tick. The deer tick is the species that can potentially spread Lyme disease. Though these tiny ticks have mainly been found in the United States, a number of infected deer ticks have been found in Manitoba. Lyme disease is treatable if diagnosed early. A vaccine has been developed.

Poison ivy, a low erect plant, grows throughout much of Manitoba. It should be avoided, as all parts of the plant contain a powerful skin irritant. It has three bright green leaflets that turn red in fall, giving rise to the refrain "leaves of three – let it be".

Winnipeg Region

Winnipeg Region
Wildlife in River City

Wildlife and cities don't mix particularly well, though many animals – usually small ones – manage to adapt to a world of concrete, streets and sidewalks, rushing automobiles and legions of two-legged creatures. In the midst of such an urban environment, parks and green spaces breathe nature back into a community.

Winnipeg's founders and their successors were remarkably far-sighted; the result is a city with an abundance of parks and places where even large animals have found a home. Assiniboine Park and adjacent Assiniboine Forest, Fort Whyte Centre and the Living Prairie Museum (among many others) provide havens for wildlife and people alike, right within the city limits.

At first glance, Winnipeg might seem little more than the gateway to an endless plain, but in fact this is "River City", with four rivers and a multitude of creeks winding through every quarter. Along these ribbons of water, riparian woodlands provide habitat for many species of birds and mammals as well as resting and staging places for many more.

Just beyond the city's boundaries are woods and water, meadows and marshes. Oak Hammock Marsh, a restored tapestry of wetlands, meadows, tall grass prairie and oak bluffs, is just north of Winnipeg. Northeast of the city, the diverse habitats of Birds Hill Provincial Park, which straddles a glacial esker, provide homes for many other species.

Nine thousand years ago, a blink of the eye in geohistory, the Winnipeg region rested at the bottom of the world's largest glacial lake. The legacy of Lake Agassiz is some of the richest soil in Canada,

Peregrine falcons, opposite, still on Canada's endangered species list, nest annually in downtown Winnipeg.

Winnipeg's rivers and streams are among its many attractions, providing picturesque and often peaceful views of the city, as well as a wealth of opportunities to view wildlife.

Ian Ward

Winnipeg Region

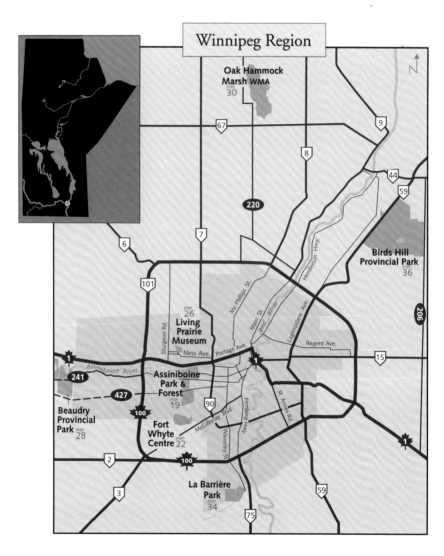

lacustrine (lake-bottom) sediment that for millennia supported the northernmost expanse of North America's tall grass prairie and the millions of bison that grazed it. Eventually, the rich land supported agriculture and commerce.

No Manitoba landscape has been more altered by human undertaking than that of the Red River Valley. Nothing can push a natural environment to the margins like a city. Yet forethought and measured growth have saved tracts of natural lands within and outside of Manitoba's capital and these remnant habitats draw a remarkable variety of life.

Trip Planning

If you have a day in the region . . . ◎ Spend some time at Oak Hammock Marsh and visit the conservation centre. ◎ Walk or cycle in Assiniboine Park and Forest. ◎ Travelling with children? Visit Oak Hammock Marsh, Birds Hill Provincial Park or Fort Whyte Centre. ◎ In winter, hike, ski or snowshoe the trails at La Barrière Park or Beaudry Provincial Park.

If you have two or three days . . . ◎ Take in the Living Prairie Museum and Oak Hammock Marsh. The next day, spend the early hours in Assiniboine Park, then hike or bike at either La Barrière or Beaudry Park in the afternoon. ◎ Taking children along? Visit Assiniboine Park (a miniature train and highly regarded zoo are located in the park). Then take in one or more of Living Prairie Museum, Oak Hammock Marsh (children love the Richardson's ground squirrels) or Fort Whyte (to get close to waterfowl and fish, as well as deer).

If you have a week or more . . . ◎ You could easily visit all seven sites featured in this region. Allow a half-day for each of Fort Whyte Centre, Living Prairie Museum, Assiniboine Park and Forest, Beaudry Provincial Park and La Barrière Park. Allow a full day or more for both Oak Hammock Marsh and Birds Hill Provincial Park.

When trip planning, keep in mind the interests of your group. A family with children might spend a week visiting sites, such as Birds Hill Provincial Park, below, that a busload of eager birders could see in two or three days.

Winnipeg Region

Assiniboine Park & Forest

Inspired by a desire to transform a sprawling railway town into a green city, Winnipeg created the city's first and largest park, Assiniboine Park, in 1909. Located in southwest Winnipeg along the south bank of the Assiniboine River and designed in the English landscape style, Assiniboine Park and adjacent Assiniboine Forest offer an accessible urban respite in a mix of landscaped and natural settings.

Assiniboine Park comprises almost 160 hectares of wooded lawns and features picnic sites, an English garden, formal and sculpture gardens, a conservatory and zoo, along with a pavilion that has been renovated to house an art gallery and restaurant. There are also playing fields for cricket, field hockey and other sports, as well as a miniature train.

In winter, visitors may enjoy cruising down a toboggan slide, skating on the duck pond, or skiing along trails. In spring and summer, people stroll through an English garden filled with flowers. Assiniboine Park Conservatory, the oldest in Western Canada, features tropical trees and exotic plants, as does the Tropical House in the Assiniboine Park Zoo.

Adjacent Assiniboine Forest is a 280-hectare sanctuary with more than three kilometres of trails through dense oak and aspen forest. A paved nature trail leads to an observation mound overlooking Eve Werier Memorial Pond.

Assiniboine Park is a haven for urban birdwatchers, and one of the best places to view a wide variety of species during spring and fall migration (typically mid-April to late May and late August to early October). The river and riverbottom forest attract migrating birds, making Assiniboine Park an excellent and accessible place to view warbler migration. Almost all of Manitoba's warblers and vireos, from the abundant yellow-rumped warbler to the

White-tailed deer, which number in the hundreds, and eastern screech-owls, uncommon in Manitoba, are among dozens of species that find a home in the varied habitats of Assiniboine Park and Forest.

Dennis Fast

Winnipeg Region

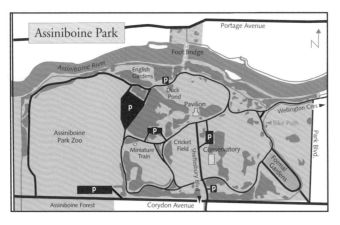

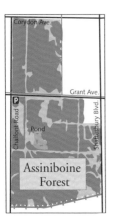

Among the aspen in Assiniboine Forest, autumn's golden grasses beckon wildlife watchers. This less-developed extension of Assiniboine Park is also a good place to see many species of birds during spring and fall migration.

uncommon Connecticut warbler can be seen here. The peak viewing time is late May. A good place to begin a birdwatching stroll is in the English Garden, where in April and early May, over a dozen species of sparrows may be seen. Common in spring and summer are American goldfinches, least flycatchers, Baltimore orioles, and eastern kingbirds. The mewing call of gray catbirds can also be heard. Cliff swallows build their tube-shaped nests under the footbridge to Portage Avenue and chimney swifts can often be seen darting after insects along the river. Wood ducks and mallards are common along the river and on the duck pond.

Cooper's hawks, fast and powerful hunters, uncommon over most of their range, have been seen nesting in the park. Eastern screech-owls, "eared" owls that are well known in Winnipeg but uncommon in the rest of Manitoba, also nest here.

Migrant warblers and vireos return in August and early September and this is also an excellent time to see ruby-throated hummingbirds in the English Garden.

In September, robins gather in large numbers prior to migration and in September and October, great blue herons and double-crested cormorants can be seen on rocks and islands in the river.

Visitors to Assiniboine Forest, particularly in spring and summer, are likely to see numerous songbirds and have a good chance of seeing white-tailed deer (which also inhabit the park) or red foxes.

Viewing images of wildlife indoors is possible at the Pavilion, left, which features a collection of paintings by celebrated wildlife artist Clarence Tillenius, as well as landscapes by W. J. Phillips and Ivan Eyre.

Photographer's Notes

June 2nd: "A short walk leads through almost all the ecosystems in the area – aspen groves, oak forest, wetlands with grassy bits here and there. Flowers everywhere – yellow lady's slipper, sweet vetches, buttercups. Deer tracks in several places along the path."

Assiniboine Park is six kilometres west of downtown Winnipeg. From downtown, drive west on Portage Avenue to Maryland Street. Turn left and continue several blocks until you cross a bridge, then right onto Wellington Crescent, and continue until you reach the east gate. Two entrances on Corydon Avenue also lead into the park. Walkers and cyclists can enter from Portage Avenue over a pedestrian bridge at Overdale Street.

Assiniboine Forest lies south of Assiniboine Park on both sides of Grant Avenue between Shaftesbury Boulevard and Chalfont Road. To enter the forest from the northwest, park in the lot on the corner of Chalfont Road and Grant Avenue.

Fort Whyte Centre

Guy Fontaine

Red foxes, like this cub, are active year-round, but are best seen in early morning and late afternoon.

Fort Whyte Centre is a non-profit organization dedicated to environmental education. Fort Whyte is named for William Whyte, one of Canadian Pacific Railway's top officials, who headed a crew of 200 men in an 1888 standoff with a competing railway over a disputed rail crossing near here.

Today, Fort Whyte is better known for its importance to wildlife. The motivation for the preserve came during the 1950s from workers at a nearby cement plant. Inspired by white-tailed deer in the surrounding woodlands and geese that took to the waters in the plant's four clay pits, the workers released three pairs of Canada geese. Now, more than 50 pairs of Canada geese nest here and the centre is operated by the Manitoba Wildlife Foundation. Thousands of geese, ducks and songbirds fly through during spring and fall migration and numerous other birds, deer and smaller mammals make it their home year-round.

The centre's self-guiding trails provide an excellent place for family excursions. Activities, films and interpretive programs teach both adults and children about our natural environment, with an emphasis on incorporating nature into urban life. Featured courses include everything from canoeing to composting. Ice fishing, snowshoe rentals, sleigh rides, a skating rink and toboggan chute make it a fun place to get outdoors in winter too.

Besides fascinating exhibits that feature live fish, as well as the Children's Touch Museum, the centre has a theatre, nature library and gift shop. In winter, visitors can view the resident flock of waterfowl in its winter home inside the centre. A number of wetland habitats are traversed by trails in the Waterfowl Gardens. Boardwalks float across marshes, bringing visitors face to face with wildlife. Trails lead through an aspen-oak forest that attracts songbirds and woodland species.

Duckweed, above, is a group name for edible (if you're a duck) free-floating stemless plants that, along with insects, will aid ducklings in reaching full size just weeks after they're hatched.

Mallards, such as this handsome drake, begin to arrive in Manitoba as early as the last week in March, and often stay until November. However, a few can be seen year-round, in Fort Whyte Centre's indoor pond.

Winnipeg Region

Boardwalks float over marshes, above, allowing visitors to the centre, including dozens of school classes, a close look at the world of wetlands.

White-tailed deer and red foxes may be seen year-round. Mallards, wood ducks and Canada geese are the most common waterfowl that nest here. During May and June, ruddy ducks can be seen courting, slapping the water with their sky-blue bills. During migration – usually late March to mid-May and late August to early October – thousands of geese and ducks stop by. Visitors can stroll along the trails among the geese. Be cautious, however. Geese are often taller than a two-year-old and can be aggressive during the nesting season.

Because the resident waterfowl are used to visitors, Fort Whyte is a great place for children to see broods of young ducklings in summer and get close to species such as hooded mergansers and common goldeneye.

Along with an abundance of waterfowl, colorful warblers flutter through the forest during migration, peaking mid- to late May. Yellow warblers are common and nest here each spring. You may also hear the rich, melodic song of a rose-breasted grosbeak, which one ornithologist described as "similar to that of a robin that has taken voice lessons". American goldfinches – often mistaken for wild canaries because of their bright yellow coloring – are common at bird feeders in the forest. Goldfinches are one of the latest nesting species, waiting until August when thistle seed, their primary food, ripens.

After the snow flies, watch for white-breasted nuthatches and listen for the *rat-tat-tat* of hairy and downy woodpeckers. Depending on the availability of food in the Interlake and other areas of the province, many northern birds – redpolls, pine siskins, pine grosbeaks – can be common visitors one winter and absent the next.

Fort Whyte Centre's Aquarium of the Prairies, the largest facility of its kind in Western Canada, features freshwater fish in two large tanks holding more than 80,000 litres of water. They depict two distinct aquatic habitats. Huge channel catfish, goldeye and rock bass swim in the smaller clay and rock "Red River" tank; smallmouth bass, walleye and sturgeon explore among the boulders and deadfall in the "Lakes of the Canadian Shield" tank. Visitors may poke their heads into a plexibubble to discover what it's like to be among the fish.

Guy Fontaine

Access Fort Whyte is 20 minutes southwest of downtown Winnipeg. Follow McGillivray Boulevard west and turn north on McCreary Road. The centre is open daily, year-round. Admission is charged. An all-terrain wheelchair is available.

Many of the geese, left, that stop at Fort Whyte in spring and autumn spend their summers farther north, but Canada geese are known to nest all over Manitoba, including its towns and cities.

From here, most go south along the Mississippi Flyway, stopping along that route to winter in Minnesota, Missouri, Arkansas and southwest Louisiana.

Geese travelling through western Manitoba are known to winter as far south as Texas.

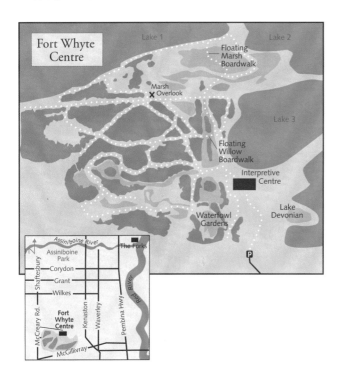

Winnipeg Region

Living Prairie Museum

"We looked out and beheld a sea of green, sprinkled with yellow, red, lilac, and white ... As you cannot know what the ocean is without seeing it, neither can you imagine the prairie."
George M. Grant, 1873

Throughout spring and summer a different mix of yellow, blue and purple flowers can be seen every week along the self-guided trails at the Living Prairie Museum in northwest Winnipeg.

Today, most people may think of prairie as waving fields of grain, but the original prairie in this region is a rich and complex ecosystem of tall native grasses and flowers that once covered 6,000 square kilometres of the Red River Valley. Some of these plants, now very rare, can be seen here in this outdoor museum, either on your own or on guided hikes with naturalists.

Consisting of an interpretive centre and mowed trails situated on 12 hectares of tall grass prairie, the museum features a habitat that once extended from Texas to southern Manitoba. (Information on a larger remnant, the 2,200-hectare Manitoba Tall Grass Prairie Preserve in southern Manitoba, can be found on page 125.) Together, these remnants sustain less than one percent of what was once a vast sea of tall grass prairie, now one of the most endangered ecosystems on Earth.

The preservation of tall grass prairie in Manitoba is particularly important, for it is one of only two provinces in Canada

Manitoba's provincial flower, the prairie crocus, above, symbolizes spring. Big bluestem, at right, is the characteristic plant of the tall grass prairies.

(Ontario is the other) where this endangered habitat occurs.

Many of the more than 200 species of native plants and wildflowers are small, low-growing plants, such as crowfoot violet, but others, such as giant hyssop, prairie lilies, roses and clovers are much more showy. Prairie plants mature and flower throughout the spring and summer.

Purple prairie clover has deep lavender flowers in June and July. Culver's root is a rare beauty with multiple candelabra spikes of white that flowers in July and August. Butterflies and hummingbirds hover near metre-high bergamot with its masses of purple blooms. The grasses are no less magnificent. Big bluestem grows nearly two metres high.

In addition to the butterflies that flit among the wildflowers, a number of typical grassland bird species may be seen and heard, including western meadowlarks, clay-colored sparrows and vesper sparrows. Residents of the aspen forest edge on the north boundary of the museum include Baltimore orioles, great crested flycatchers (with their long rusty tails and crewcut headcrests), yellow warblers, gray catbirds, warbling vireos and black-capped chickadees. But the main draw here is the tall grass prairie habitat.

The leaves of bergamot, left, and raspberry were used in a tea by settlers in Manitoba.

1 c. fresh raspberry leaves
1 c. fresh bergamot leaves
4 c. boiling water

Pour boiling water over the leaves, cover and steep for 10 minutes. Strain and bring to a boil.

Two tablespoons each of lemon juice and honey may be added.

Access: The Living Prairie Museum is situated in northwest Winnipeg at 2795 Ness Avenue. There is an observation deck on the second level of the museum. The interpretive centre is open weekends from mid-April to the end of June, daily from July 1st to the first weekend in September, and by appointment the rest of the year.

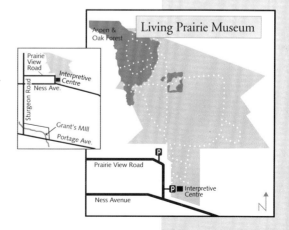

Winnipeg Region

Beaudry Provincial Park

Despite its iridescent breeding plumage, it is often the long and varied song of the indigo bunting, heard in June and July, that gives its presence away.

Straddling the Assiniboine River, Beaudry Provincial Park offers a natural respite just minutes west of Winnipeg. Even on a typically warm summer day, visitors will find plenty of space to enjoy the woodland and meadows.

Combined, the forest and prairie total 1,200 hectares, providing a year-round oasis surrounded by farmland for strolling or skiing. Numerous unmarked trails intersect the two main trails on the south side of the winding river: the 2.5-kilometre Wild Grape Trail and the 2.8-kilometre Elm Trail. In winter, additional cross-country ski trails on the north side of the river make the park a great destination.

The dense riverbottom forest along the meandering Assiniboine River is home to some of the largest cottonwood, Manitoba maple and basswood trees in the province. One old cottonwood measures 5.8 metres around. The vines of wild grapes, bearing small, tart fruit in the fall, climb the shrubs and trees; saskatoons, chokecherries and hazelnuts abound. The trails wind through a forest floor covered in lush ferns, but take care if you leave the paths, for poison ivy and stinging nettle are found here too.

The park is also dedicated to restoring remnants of tall grass prairie. It's a

Beaudry's lush riverbottom forest, opposite, supports many species of birds and mammals.

process that may take generations, but is particularly worthwhile, since this is one of the most endangered habitats in North America.

Songbirds, woodpeckers and white-tailed deer are common here, and beaver, muskrats, raccoons, foxes, woodchucks and great horned owls may also be seen. During May and June visitors have an excellent chance of seeing or hearing red-headed woodpeckers, yellow-throated vireos, warbling vireos and rose-breasted grosbeaks. Black-and-white warblers can be seen making their way up and down old tree trunks in their search for insects. There are also great crested flycatchers, wood ducks and catbirds.

Adding color are yellow warblers and bright yellow American goldfinches that gather in sociable groups to feed on thistle seeds or bathe in puddles after a rain. Less easily seen, but usually resident here, are brilliant blue indigo buntings, as well as scarlet tanagers, the males with their distinctive red bodies, black wings and tails, and the yellow-bodied females.

The rapid *coo-coo, coo-coo-coo* of black-billed cuckoos can sometimes be heard as they hop from branch to branch in search of hairy caterpillars. At night, listen for the barred owl's *Who cooks for you? Who cooks for you all?*

In mid-August, small white mayflies hatch from the river in the evening in such numbers that it appears to be snowing. If you stand by the river you can hardly see the far bank.

Beaudry Park is 11 kilometres west of the Perimeter Highway on Roblin Boulevard (PR 241). Designated trails are accessible from a trailhead at the parking lot off PR 241. Non-modern washrooms and a canoe launch are provided.

Jerry Kautz

Barbara Endres

At home in water, raccoons can sometimes be seen along the banks of the Assiniboine, searching for frogs and clams beneath the water with their dextrous forepaws. Much like human hands, these five fingered paws leave distinctive tracks along the riverbank. Raccoons are "pacers", moving both limbs on one side of their bodies, then both limbs on the other side. This leaves paired tracks with a hind foot beside a front foot.

Jerry Kautz

Oak Hammock Marsh WMA

Guy Fontaine

Silhouetted against the evening sky, these geese join the hundreds of thousands that gather each spring and fall at Oak Hammock Marsh Wildlife Management Area.

If not for a determined rescue effort, Oak Hammock Marsh, 30 kilometres north of Winnipeg, might well be gone. More than 100 years ago, it was a massive marsh, called St. Andrew's Bog by early settlers. A small grove of oaks near the marsh attracted picnickers from Lower Fort Garry. Through the 20th century, however, drainage efforts succeeded in reducing this vital wetland to less than one-tenth of its original size.

Today, substantially restored, Oak Hammock Marsh Wildlife Management Area provides one of the best migratory bird viewing areas in North America. The WMA includes the Oak Hammock Marsh Conservation Centre, which houses an interpretive centre and serves as the headquarters for Ducks Unlimited Canada. Thousands of people visit the centre each year to walk the boardwalks and trails, experiencing marsh life up close and personal. During peak migration in April and October, the sound of Canada geese fills the air as they stop to rest and feed. Skeins of geese are often visible as far as the eye can see.

Encompassing an area of roughly 3,500 hectares, the marsh is peppered with more than 50 islands designed to provide safe nesting sites for marsh birds. Edged by cattails and bulrushes and dotted with willow bluffs on the upland areas, the marsh provides ample nesting cover for dozens of species of waterfowl and shorebirds. A significant remnant of tall grass prairie, the Brennan Prairie, is on the west side of the WMA, attracting sharp-tailed grouse and other grassland birds.

Just northwest of the interpretive centre on Peregrine Drive, water from an artesian well bubbles from the ground into a water supply canal. Because the spring flows year-round, this is a great place to visit after the snow falls, to see snow buntings and redpolls or to capture interesting ice formations on film. Snowy owls may be seen on the snow-covered marshes, or on utility poles on the drive in.

More than 280 species of birds have been recorded at Oak Hammock's bountiful (and carefully managed) wetlands, but there is room for other animals as well, such as this Richardson's ground squirrel.

One of the principal goals of the Oak Hammock Marsh Conservation Centre, right, is public education about the importance of preserving and enhancing wetland habitats so that waterfowl, opposite right, can continue to thrive.

At the periphery of the marsh, lure crops are planted to prevent the migrating masses from damaging farmers' crops in surrounding fields. Some local landowners also use propane bangers or invite hunters to assist in damage control.

Oak Hammock Marsh teems with wildlife. An estimated half-million birds stage here spring and fall, with peak numbers usually mid-April to mid-May and again in the latter part of September. More than 280 species have been recorded, along with 25 mammalian species and several amphibians.

In terms of sheer numbers, Canada geese are the marsh's top visitors, but birders can also expect to see snow geese, a dozen species of ducks such as mallards and blue-winged teal, as well as several grebe species. Ruddy ducks, their chests puffed, court their mates, while pied-billed, eared and horned grebes nest in floating vegetation.

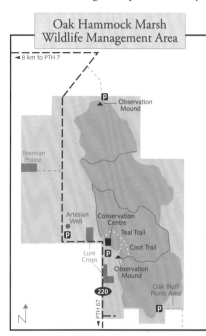

When water levels are reduced in one or two cells in mid-May or July and August, this can be one of the best places to view a variety of shorebirds feeding on exposed mud flats, such as greater and lesser yellowlegs, American avocets, and semipalmated, Baird's and white-rumped sandpipers.

Wilson's phalaropes, cousins of sandpipers, are also here. Phalaropes, with the exception of the Wilson's, are rare inland of Canada's northern coasts. The female is larger and more colorful than the male; during the summer she can be identified by her rust-colored shoulders and neck. If you visit just before dusk, you might hear an

American bittern, though spotting this elusive and well-camouflaged wader takes patience. During May and June, its hollow "pumping" call resonates through the twilight hours: *oonck-a-tsoonck, oonck-a-tsoonck*.

Though outnumbered by waterfowl, migrating songbirds also move through Oak Hammock in spring and fall. In October, rough-legged hawks perch on fenceposts by the roadsides, or hover over the nearby fields in search of mice. And many species that are rare in Manitoba, such as cinnamon teal and cattle egrets, have been recorded at Oak Hammock.

Mammals are also very much at home here and can be readily seen. A 200-strong colony of Richardson's ground squirrels lives in two mounds near the centre. Jackrabbits bound across the uplands in summer and have been seen on the pond ice in winter. Muskrats are abundant, and white-tailed deer raise their young in the wooded areas adjacent to the wetlands.

In 1985, it was estimated that 70 percent of wetlands in North America had been drained since European settlement. A year later, Canadian and American governments initiated the North American Waterfowl Management Plan, aimed at restoring waterfowl populations to levels experienced in the 1970s.

Access From Winnipeg, follow PTH 7 north of the Perimeter Highway (PTH 101) 18 kilometres to PTH 67, turn east and continue for eight kilometres, then north four kilometres on PR 220 to the main parking lot at the western edge of the marsh. The centre can also be accessed from PTH 8, by turning west onto PTH 67 for seven kilometres to PR 220. The WMA offers viewing mounds, boardwalks and more than 20 kilometres of dike trails. Oak Hammock Marsh Conservation Centre, a joint project of Manitoba Natural Resources and Ducks Unlimited Canada, features marshland wildlife displays, interpretive programs, a 120-seat multimedia theatre, meeting rooms, observation decks, washrooms, a cafeteria and gift shop. The centre charges an admission fee, which includes access to all facilities and participation in guided tours. A bird checklist is available at the centre gift shop.

La Barrière Park

This popular three-square-kilometre city park includes a portion of the La Salle River just south of Winnipeg. Unmarked but well-maintained trails for walking, biking and cross-country skiing meander along the river and cross it via a walking bridge. The slow, winding river, some 20 metres across, is an ideal place for novice canoeists.

A broad band of bur oak, Manitoba maple and aspen lines both banks of the river. On the south side, lawns and wild meadows stretch back from the woods. During summer, saskatoon bushes provide food for birds and small mammals, and in the late summer, chokecherry and hazelnut bushes and wild plum trees are often thick with fruit.

White-tailed deer are often seen here by cross-country skiers using the riverside trails in the winter and by canoeists in the summer. In spring and summer, riverbank foliage makes them more difficult to spot, but quiet paddlers may see a doe and fawn coming down to drink. Beavers and muskrats can also be seen along the river, while wood frogs populate the riverbanks and seasonal ponds in the park.

Songbirds, raptors and woodpeckers are the main types of bird life in the park. This is an excellent place to view scarlet tanagers, which nest in the park almost every spring. In early morning and evening, red-headed woodpeckers, yellow-throated vireos and lark sparrows may be heard, along with the melodious whistles of the orchard oriole. At midday, picnickers may be serenaded by an indigo bunting. The male, cloaked in blue breeding

Common to almost all of North America, great horned owls are year-round residents here, preying on small mammals and birds.

Wood frogs, found in much of Manitoba, prefer a woodland environment to open meadows. Their barking mating call (sometimes likened to a duck quacking) can be heard as early as mid-April.

North America's only red bird with black wings and tail, scarlet tanagers lose their brilliant breeding plumage in late July and August, when the red feathers are replaced by greenish-yellow plumage.

plumage, sings his mating song of paired notes from the tops of trees. Baltimore orioles, which build pendulous nests in the upper forest canopy, can sometimes be seen, their bright plumage a flash of fire against the trees. At dusk or during the night, great horned owls may be heard.

Access The park is located on Waverley Street, six kilometres south of the intersection with the Perimeter Highway (PTH 100). Immediately after crossing the bridge over the La Salle, turn west to the parking lot. The area on the west includes the open picnic area, canoe launch, forest trails and the walking bridge. A second lot to the east of Waverley is used mostly during the winter by cross-country skiers.

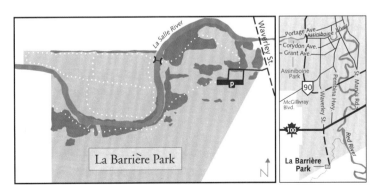

Birds Hill Provincial Park

Black-capped chickadees, year-round residents here, are friendly and easily attracted to feeding stations.

Located on a glacial esker, Birds Hill Provincial Park rises above the surrounding lacustrine plains of the Red River Valley. In the 1800s, long before a floodway existed to divert spring meltwater around Winnipeg, people came here during floods to camp on the hills above the rising water.

Today, Birds Hill is known for deer and wildflowers, bird life and the Winnipeg Folk Festival, held every July. The park's 3,350 hectares support oak and aspen forests, a cedar bog, black spruce stands and grasslands bright with native prairie flowers.

The park bustles with activity in the summer as people come out to camp, walk, cycle, swim or horseback ride. Four self-guiding trails beckon walkers. Three – the Bur Oak, Cedar Bog and White-tailed Deer Trails – are covered in more detail in this section. A paved 7.2-kilometre trail through the predominantly aspen parkland and around the lake attracts cyclists and in-line skaters. Other trails are designated for cyclists, hikers, horseback riders and horse-drawn vehicles. In winter, add snowshoeing, cross-country skiing, dog sledding and even snowmobiling to the mix.

Most of Birds Hill is aspen-oak parkland. From spring to fall, grasses and wildflowers continually change the color of the meadows. In late May and early June, several dry prairie ridges at Birds Hill turn pinky purple with three-flowered avens, also known as prairie smoke. When the flowers go to seed, they appear to cover the fields with a grayish purple "smoke." In summer, the yellow hoary puccoon and later, purple wild bergamot and large yellow lady's slippers are common in the grasslands, while showy lady's slippers can be found in wooded areas.

Watch, too, for showy milkweed, necessary to the life cycle of monarch butterflies, blue-eyed grass with its dainty blue flower, and giant hyssop, used to make a mint-like tea. Rarer species include western silvery aster and whorled milkwort.

Birds Hill's trails offer excellent opportunities to view wildlife close to Winnipeg.

One of Manitoba's highest concentrations of white-tailed deer lives in the park. Beavers, eastern chipmunks, red foxes, muskrats, white-tailed jackrabbits, snowshoe hares, striped skunks, red and gray squirrels and Richardson's ground squirrels are common mammals. Boreal chorus frogs, gray tree frogs, northern leopard frogs, northern spring peepers, wood frogs and American toads provide evening entertainment.

The park was named after the Bird family, which owned land in the vicinity, but one visit will convince you it was named for the birds. More than 200 species of prairie, parkland and boreal forest birds have been seen here. During summer, the varied song of brilliant blue indigo buntings may be heard from the tops of aspens and oaks, and the low-pitched buzzing of clay-colored sparrows is common along the trails. Eastern towhees have been seen in the brushy areas near the west end of the loop drive and Manitoba's provincial bird, the great gray owl, is occasionally seen in the tamarack-black spruce bogs along the eastern edge of the park.

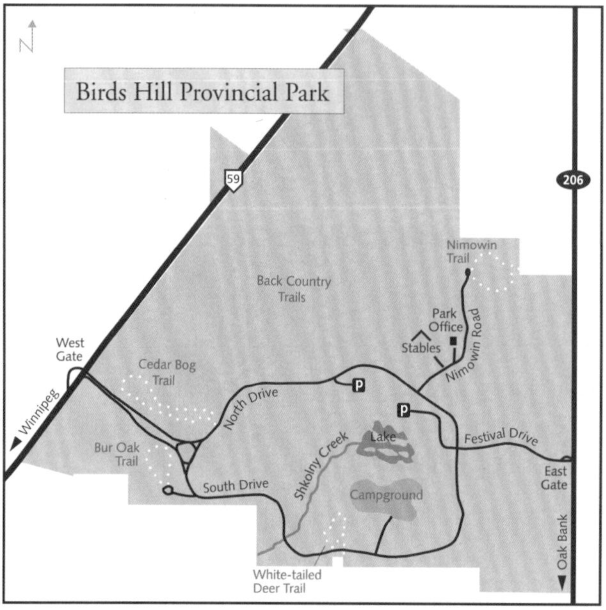

Flocks of juncos and a variety of warblers are common in the fall. Look also for northern goshawks and northern shrikes. In winter, brilliant yellow evening grosbeaks and strawberry-red pine grosbeaks brighten the landscape.

From Winnipeg, follow PTH 59 north 19 kilometres from the Perimeter Highway (PTH 101) to the west gate entrance. North and South Drives form a 10-kilometre circular route inside the park. Maps and information can be obtained during the summer months from the park office and at most trailheads. Facilities include camping and picnic sites, a riding stable and a restaurant.

Bur Oak Trail

This paved one-kilometre self-guiding trail loops through a stand of gnarled bur oaks, offering an easy and wheelchair accessible outing from spring to fall. Beginning at a parking lot adjacent to a grassy picnic area, the trail leads north. Along the way, seasonal interpretive signs provide information on the plants and animals of this habitat. About halfway along the paved loop, a sign marks a rugged and often wet three-kilometre extension that leads through a former gravel pit, past tall spruce and poplar trees, then climbs to drier grasslands before returning to the paved loop.

In late April and early May, you may find prairie crocus and early blooming violets along the way. Later in May, the trail is brightened with flowering shrubs. Nine shrubs with edible berries grow in the park, including chokecherry, highbush cranberry, nannyberry, saskatoon and wild plum. These, and hazelnuts, draw birds and small mammals to the area. In late June, Wood's roses scent the air. Berries follow in succession over the summer months, finishing with highbush cranberries in September.

In early summer, visitors are almost guaranteed to see indigo buntings on this trail, often singing from a dead branch at the top of a stunted bur oak. Other birds include clay-colored and chipping sparrows, red-eyed and warbling vireos, least flycatchers, downy woodpeckers and Baltimore orioles. Rarer, but sometimes seen, are lark sparrows and black-billed cuckoos. Looking up, one may see red-tailed, broad-winged or Swainson's hawks soaring. On early mornings or evenings in May or early June, ruffed grouse can sometimes be heard drumming.

Downy woodpeckers, such as this male, are similar in appearance to hairy woodpeckers, but smaller. Their call is a soft pik, pik, pik.

Access — Enter the park from PTH 59, turn on to South Drive and take the first road right marked with a brown hiking sign. The level trail is wheelchair accessible. Braille text for the trail brochure is available in the park office.

In late May, the trail is awash with blossoms such as these, which promise chokecherries at the end of August.

Cedar Bog Trail

One of Birds Hill Park's most unusual trails, this all-season 3.5-kilometre self-guiding trail leads through a mixture of grasses, trembling aspen and bur oak to a magnificent stand of eastern white cedar. White cedar habitat is more common to the Great Lakes-St. Lawrence forest region of Ontario and Quebec, which extends into extreme southeastern Manitoba. The tall cedars form a canopy that blocks sunlight and creates mysterious shadows along the trail through the cedar bog. In early to mid-June, notable plants include small yellow lady's slipper, Labrador tea and highbush cranberry, which blooms in clustered masses of white flowers.

In winter, the trail is hard packed for walking, but snowshoes may be required after a storm.

Red squirrels and Franklin's ground squirrels can be found in the aspen-oak parkland and the cedar forest, where they feed on acorns, cedar cones, hazelnuts and the fruits of many shrubs. Forest insects provide food for downy, hairy and pileated woodpeckers. Look for lark sparrows near the parking area and along other forest edges.

Food is less plentiful in the cedar bog than it is in the parkland habitats and both mammals and birds are fewer in number, but yellow-rumped warblers, with their yellow crown patch, yellow wing markings and soft, warbling song, can sometimes be seen or heard among the conifers. The Connecticut warbler, a relatively rare warbler, has been seen here in the spring. During April and May, woodcock can be found in the low wet areas. Mainly nocturnal, these stocky, well-camouflaged birds often allow a close approach before exploding into flight with whistling wings. Their call is a nasal *peent*.

Unlike their white-breasted cousins, which are common in Winnipeg's largely deciduous urban forest, red-breasted nuthatches prefer conifers. They share the family acrobatic abilities, however, and can often be seen walking straight up or down tree trunks, or hanging and feeding upside down.

In winter, white-tailed deer are often seen, along with snowshoe hares, sometimes with red foxes in pursuit. Blue jays and pine grosbeaks are here, as well as boreal chickadees and red-breasted nuthatches. In very cold weather, the ruffed grouse will plunge into deep snow to create a burrow, known as a kieppe. When surprised, it may burst from its roost in an explosion of snow.

The twining roots and trunks of some of Manitoba's largest eastern white cedars, below, create both picturesque patterns for visitors and appropriate abodes for residents like this nesting woodcock.

Access Enter the park from PTH 59 and turn on to North Drive. The first road to the left or west, marked with a brown hiking sign, leads to the trailhead and parking lot.

Winnipeg Region

White-tailed Deer Trail

Graceful and numerous, white-tailed deer are relative newcomers to Manitoba. Vast herds of bison, along with smaller numbers of mule deer and elk dominated the central plains for millennia; their grazing habits and regular prairie fires fostered grasslands and kept the woodlands at bay. The disappearance of the great herds in the past 150 years and the suppression of fires that accompanied settlement changed all that, turning grassland into aspen parkland in many places in Manitoba and creating ideal habitat for white-tailed deer.

The 1.5-kilometre White-tailed Deer Self-guiding Trail meanders through aspen-oak forest and across large, bowl-shaped meadows. Interpretive signs explain the biology, behavior and management of the deer in the park. Near dawn or dusk, a quiet pause in the viewing tower located midway along the trail may reward viewers with a glimpse of animals feeding in the meadow below. Dainty hoof prints on a section of trail prove that the deer are nearby.

The trail also offers opportunities to view wildlife other than deer. At the trailhead, jackrabbits can often be seen. Eastern chipmunks and red squirrels live here too, feeding on acorns and plentiful berries.

Birds seen along the trail in spring and summer include yellow-bellied sapsuckers, softly tapping their rows of holes into living trees, least flycatchers, tree swallows, red-eyed vireos, American redstarts and ovenbirds, clay-colored, chipping and lark sparrows, Baltimore orioles and American goldfinches. Here and in several other areas of the park, wild turkeys are often seen.

The trailhead is just west of the campground road off South Drive.

The dainty tracks of white-tailed deer, among Manitoba's most easily-seen mammals, can often be observed along this and other trails at Birds Hill Park in any month of the year.

Red squirrels, also abundant through much of the province, stay in their nests during the coldest weather, but can often be seen playing among the branches on sunny winter days.

Eastern Manitoba

Eastern Manitoba
Shield & Forest

Once, when the world was young, eastern Manitoba was alpine, a region of jagged peaks and deep chasms. More than two billion years later, the ancient mountains have been eroded to a nub. But what a nub it is. Today, we know this primeval bedrock as the Precambrian Shield, a foundation that serves as the country's geological backbone, stretching from Newfoundland to Yukon.

This landscape has become synonymous with the image of Canada itself, an endless expanse of clear lakes, rocky shores and boreal forests. Today, though the rocks and the forests of Manitoba's shield region have provided minerals and timber for more than a century, in many places wilderness remains, protected by provincial designation, as in Atikaki Provincial Wilderness Park, or the Wilderness Zone of Manitoba's most visited park, the Whiteshell.

Whiteshell, Nopiming, Atikaki: the trio of provincial parks marches up Manitoba's east side from south to north, each more remote than the last, yet each in its own way accessible to those interested in watching wildlife.

This is large mammal country. Atikaki means "country of the caribou" in Anishinaabe and woodland caribou can still be found here and in Nopiming in small isolated herds. The permanent inhabitants of all three parks include moose, black bears and wolves.

Eastern Manitoba is not only Precambrian Shield, however. Toward its western edge, it also includes ancient beach ridges and glacial moraines, legacies of the Wisconsin glaciation.

Though soils are thin in this rocky world, plant life is abundant. The boreal forest is typically black and white spruce, jack pine, tamarack and balsam fir, intermixed with trembling aspen,

A commanding presence even at rest, the great gray owl is beautiful and solitary, an apt symbol of Manitoba's wild places.

The still water of "Alf's pond" at Whiteshell's Alf Hole Goose Sanctuary attracts avian and human visitors alike.

Eastern Manitoba

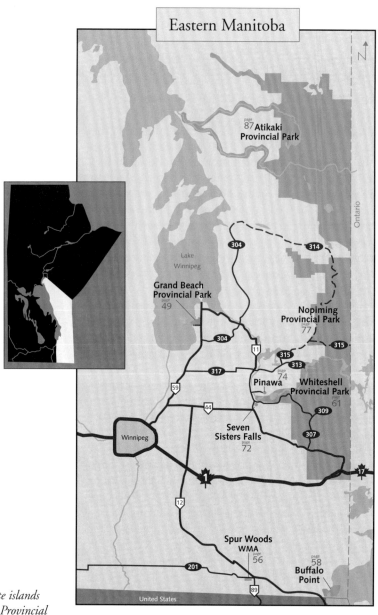

Even remote islands in Atikaki Provincial Wilderness Park, opposite right, can be reached by air or water.

black poplar and birch. Grand Beach lies to the west of the shield, sharing much of its vegetation and wildlife, but here the bedrock lies buried far beneath the surface.

Accessible, yet in many places as secluded as any part of the province can be, eastern Manitoba remains one of the most unforgettable places in the world to view wildlife.

Trip Planning

If you have a day in the region . . . ✿ Head to Grand Beach Provincial Park to hike the Ancient Beach and Wild Wings Trails or spend some time walking and birding along the beach. ✿ Drive through Whiteshell Provincial Park after an early morning birding stop at Seven Sisters Falls. Continue along PR 307 and PR 44 with selected stops at Pine Point, Bannock Point, Alf Hole Goose Sanctuary, McGillivray Falls Trail, White Pine Trail or the Lily Pond.

If you have two or three days . . . ✿ Find a great camping spot or resort in the Whiteshell and allow more time to hike and bike the trails or paddle the lakes. Add birding in Pinawa to the mix. ✿ Paddlers may want to plan a short canoe trip in Nopiming Provincial Park along the Bird River. ✿ Drive the Northwest Angle loop (which, despite its name, is Manitoba's most southeasterly corner) heading east on the Trans-Canada Highway, then south on PTH 12 to Buffalo Point and Moose Lake, with a stop at Spur Woods. On your return via PR 201 and PTH 59, spend a half-day at the Tall Grass Prairie Preserve (featured in the Southern Manitoba section).

If you have a week or more . . . ✿ Spend three days at Grand Beach and four in the Whiteshell. ✿ Or split your time between Nopiming and the Whiteshell. ✿ Anglers or wilderness enthusiasts: fly into Atikaki to fish at Sasaginnigak Lake, paddle the Bloodvein River or raft the Pigeon River. ✿ Hike the Mantario Trail or canoe to Mantario Lake.

Rick Hurst

Eastern Manitoba

Grand Beach Provincial Park

Grand Beach, on the eastern shore of Lake Winnipeg – the world's 13th-largest and Canada's sixth-largest freshwater lake – can justifiably claim to be a place for all seasons. Created by the long winter of the last glaciation, its white sand beaches have been a summertime lure for Manitobans for more than a century. From spring through fall, the juxtaposition of beaches, lagoon, marshes and the nearby woodlands also make Grand Beach a terrific spot for birding.

Native aspen, birch and jack pine thrive in the sandy soil of the uplands, along with red and Scots pine that have been planted there, and a profusion of berry-producing shrubs, including chokecherries, raspberries and blueberries. The extensive marsh and lagoon area behind the beaches and dunes, which encourages the growth of cattails and bulrushes, is home to a great diversity of birds and small mammals. The waters of both lake and lagoon also provide prime habitat for many types of fish.

Supper is always at hand for the osprey pair that inhabits this large and lofty nest, above, overlooking the lagoon at Grand Beach, opposite.

The park attracts many types of birds – water, shore and marsh species, as well as raptors and songbirds – including one species on Canada's endangered list and two more that have made a dramatic recovery since DDT was banned in 1972.

Undoubtedly the most significant of these are the endangered piping plovers, tiny white, beige and black shorebirds with a love for beaches that, unfortunately, matches our own (see Dunes & Beaches in this section). Grand Beach is also prime nesting habitat for ospreys and bald eagles. Both species have made a dramatic recovery in the past two decades.

Just outside the park boundary, a pair of ospreys nested for more than 15 years on the top of a hydro substation. Each year they added to their nest in preparation for a new clutch of young. By the mid-1990s, the nest was so enormous that hydro officials began to worry about the safety of the station. In response, several nesting platforms were erected atop utility poles, including one located just across the road from the substation. This platform, along with the efforts of a pair of bald eagles who helped to "evict" the ospreys, enticed the pair to a new home across the road. The bald eagles briefly moved into the enormous vacated nest, but abandoned it soon after; it was likely too busy for them. A spotting scope is provided nearby to allow visitors a rare view of the ospreys, among North America's most magnificent birds of prey.

The lagoon and its adjacent marshlands are also home to a variety of ducks and marsh birds (see Wild Wings Trail), and attract loons and mergansers heading north in the spring. On the right day in May, birders might see between 50 and 100 different species of birds, including many species of warblers, as well as flycatchers and thrushes, in the various habitats of Grand Beach.

Manitoba's prevailing westerly winds mold the fine white sand into dunes that not only provide habitat for a variety of species, but serve to protect the leeward wetlands.

Fast Facts

At 24,085 square kilometres, Lake Winnipeg is more than half the size of Switzerland.

Grand Beach is located on the east side of Lake Winnipeg about 80 kilometres north of Winnipeg. Follow PTH 59 north of the city to the well-marked junction of PTH 12. Turn west and follow the signs to the East or West Park Gates.

Ancient Beach Trail

The Ancient Beach Trail celebrates an unusual chapter of Manitoba's glacial history with a 2.1-kilometre self-guiding trail that includes not one, but two magnificent viewpoints from the top of the Belair moraine. Composed of fine white sand studded with large boulders, the moraine is quite different from the layered sand and gravel glacial remnants found elsewhere in the province.

Scientists believe it was formed not during the retreat of the Laurentide ice sheet, but during its advance. Sometime prior to 14,000 years ago, two lobes of the vast Laurentide ice sheet converged on what is now the Grand Beach area. They carried enormous quantities of very fine sand, gathered when they scraped across the bedrock of central Manitoba. This sand was deposited in glacial meltwater along the cleavage between the glacial lobes, along with thousands of large boulders the ice had transported from the Precambrian Shield to the northeast. As the two lobes moved slowly south, they created a series of sand hills embedded with large rounded rocks.

About 12,000 years ago, glacial Lake Agassiz was born along the southern edge of the retreating ice sheet. In time, it grew to be the largest lake the world has known. The hills of sand (and everything else in southeastern Manitoba) disappeared underwater for about 3,000 years, until at last the Belair moraine poked above the water, an island in a vast, cold, freshwater sea.

As the lake drained, the emerging sand hills were molded by the waves. The moraine's once-rounded profile became a series of slopes and beaches, littered with boulders that had been embedded in the sand. By 8,000 years ago, Lake Agassiz was gone. In fact, Lake Winnipeg's south basin may have been dry for a time, with the lake's southern shore about where Hecla Island is now. Then slowly, the land at the north end of the lake, relieved of its great burden of ice, rose to reflood Lake Winnipeg's south basin.

The Grand Beach uplands lie in the transition zone between the Manitoba Lowlands and the boreal forest of the Precambrian

Gregarious by nature, this male evening grosbeak enlivens winter woodlands with his bold yellow forehead, bright yellow belly and back, and resonant call, a loud, ringing cleep.

Shield. The Ancient Beach Trail displays a diversity of habitat and bird life, as it climbs from a dense deciduous forest into lovely stands of red and jack pine, birch and aspen. Blackburnian and pine warblers can be seen in the treetops and branches of the conifers, while white-throated sparrows prefer thick underbrush. Near dusk, the rising flutelike notes of a Swainson's thrush can sometimes be heard.

In winter, when the trails at Grand Beach attract cross-country skiers, a variety of boreal forest birds can be sighted, including black-capped chickadees, red-breasted nuthatches, gray jays and evening grosbeaks. Three-toed woodpeckers, the male with its bright yellow cap, can sometimes be seen along the trail and their presence can be detected by patches of bark scaled from still-standing dead jack pines.

Black bears have been sighted on the trail during blueberry season, in late July and early August.

A reminder of ages past, this large boulder was transported by ice and revealed by water. The lichen adorning it is a more recent addition, perhaps only centuries old.

The Ancient Beach Trail is located north of the East Park Gate, not far from the campground. Parking is available at the trailhead on the east side of the access road. Brochures, also available here, explain the features of the moraine as you climb to the top and descend the slope.

Dunes & Beaches

Three kilometres of white sand beaches backed by rippling dunes were what first attracted vacationers to Grand Beach. And on summer weekends, the park's exquisite strand still draws thousands of appreciative sun-lovers. But there is more to this beach than holidays. This is also ideal habitat for the endangered piping plover, a tiny shorebird on the edge of extinction.

Piping plovers have an unfortunate passion for exposed beaches. Their favorite nesting sites are on open sand, just above the high water line, open to a wide variety of predators. They are undeterred by the possibility of storms or, on Lake Winnipeg, wind-driven tides, which have been known to wash over nests and scatter the eggs. The first nesting attempt usually takes place in May. If these eggs are destroyed, the birds often respond with a second attempt. Even in fair weather, the eggs are there for the taking by gulls, crows, skunks, foxes and raccoons.

When they can, plovers position their nests safely above the storm tide mark. The adults also have a repertoire of convincing evasive tactics, including feigning injury, with dragging wings and loud peeps, that often succeed in diverting would-be predators, at least of the four-footed variety. People, alas, are less easily distracted. Beachfront development, along with recent high water levels in Manitoba, appear to have contributed to the species' decline in this province. The 1996 Manitoba census found 60 adult plovers, including just 24 breeding pairs – a significant decline from the 114 adults identified in 1994.

Plovers do best when left alone; studies have shown that breeding pairs at isolated sites rear at least twice as many young to adulthood as do birds that nest on beaches frequented by vacationers.

Other animals use the beaches too, mostly when people aren't around. An early morning or evening stroll often reveals the dainty imprints of white-tailed deer, the ambling tracks of a raccoon or even the unmistakable prints of a young black bear.

Piping plovers, above, are now protected in a number of sites in Manitoba and their known nesting sites at Grand Beach are fenced off. Park interpreters and volunteer plover guardians can provide guided viewing opportunities for visitors wishing to see the birds during nesting season.

Wild Wings Trail

This trail loops around a peninsula that juts into the Grand Beach lagoon, winds through a small grove of trees and continues along a boardwalk. Well-signed and providing excellent views over the water, it is a short (one kilometre) and generally tranquil self-guiding trail. But at certain times of the year the lagoon and adjacent wetlands resound with avian activity and are enormously popular with park visitors.

In May the lagoon teems with migrating birds, including common loons and common mergansers en route from wintering grounds in the southern United States. This is also the month of nest building, courtship and breeding for many marsh birds, and the best time to see most species attired in their breeding plumage.

In June, carp, an introduced fish species, creates the excitement. Migrating from Lake Winnipeg, the fish, which can reach weights of nine kilograms, converge through a narrow channel into the lagoon. Once there, they begin a spawning frenzy. The thrashing about, with dorsal fins flashing through the water, draws opreys and bald eagles, though the birds actually tend to prey on smaller fish such as perch.

Ospreys and eagles can be seen throughout the summer and into the autumn, along with many other smaller marsh residents, including diving and puddle ducks, western grebes, yellow-headed and red-winged blackbirds and sedge and marsh wrens. Early morning visitors should spot a great blue heron walking along the marsh edge as if in slow motion, waiting for an unsuspecting minnow to venture too close to its lightning-quick bill.

American white pelicans are also often seen in the lagoon. Not only have these large, distinctively shaped birds rebounded from their threatened status, but numbers are sufficient that they once again engage in communal feeding activities. In the lagoon at Grand Beach they can often be seen fishing together in groups, herding fish into the shallows and then scooping them, in unison, into their huge beaks. Such cooperation is unusual in the avian world.

During September and October the marsh again comes alive with migrating flocks heading south for the winter.

 The Wild Wings Trail is located midway between the East and West Gates, just west of the causeway over the lagoon.

The largest heron in North America, the great blue stands 1.2 metres tall and has a wingspan of more than two metres.

Spur Woods WMA

This small Wildlife Management Area, located within three sections of land less than 10 kilometres from the American border, was established primarily to protect stands of old-growth red pine and eastern white cedar. Several access trails, including one that follows an old railway bed, provide year-round access. Glacial beach ridges and moraines have created a rolling, primarily sandy habitat.

More than eight kilometres of wide paths pass through an array of habitats, including mature 70-year-old jack pine, deciduous forest and stands of eastern white cedar. Some of the cedars stand taller than 23 metres and are too big to put your arms around. The forest floor is covered with lichens and mosses. In June and July, visitors will often find stemless lady's slippers among them.

At Spur Woods, you may be lucky enough to see both Canada's national mammal, the beaver, and Manitoba's provincial bird, the great gray owl, as the latter swoops down on one of the plentiful red-backed voles in the woods. Here, porcupines climb trees to gnaw at the inner bark and black bears find plenty of bearberries and blueberries to eat as they fatten up for winter. White-tailed deer and moose are common. Wolves, or at least their tracks, are sometimes seen. The best place to look for them may be from the ridge on the trail at the western edge of the WMA, which overlooks a bog that reaches south to

Spur Woods provides prime habitat for this northern saw-whet owl.

Though it might seem an unlikely climber, the porcupine is designed for scaling trees. Its claws are long and curved and the first digit on each hind foot is replaced by a broad, moveable pad that allows a firm purchase on branches. Normally nocturnal in spring, summer and fall, porcupines will feed at any time, day or night, during winter.

Minnesota. Farther along, the same broad trail leads to the railway bed and, following the abandoned line west, to stands of white cedar.

A second trail branches off from the first and offers exceptional old-growth forest viewing. Because shrub growth is minimal, it is easier here to spot black bears or grouse. Spur Woods is within the major breeding and migration corridor for northern forest owls, including great gray, northern saw-whet, barred and boreal owls. Other birds include species associated with old-growth forest, such as pileated woodpeckers and gray jays.

Access Spur Woods is six kilometres west of Piney on the south side of PR 201. Unmarked trails lead south into the forest, inviting exploration, compass in hand. In winter, snowshoes or cross-country skis may be needed to travel on some of the trails.

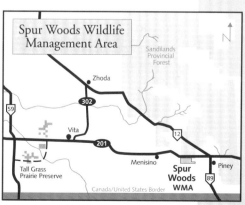

Buffalo Point,
Birch Point & Moose Lake

Buffalo Point stretches into Lake of the Woods, marking Manitoba's most southeasterly extent.

Buffalo Point, Birch Point and Moose Lake are located in the extreme southeastern corner of Manitoba. Buffalo Point juts out into Lake of the Woods on the Canada-United States border. North of it, across Buffalo Bay, is Birch Point. Moose Lake is a few kilometres inland.

Buffalo Point

Joined to the mainland by a bog, this 1,600-hectare peninsula is situated on Buffalo Point First Nation land, on southern Lake of the Woods. It includes bog, with willow and tamarack, and mixed forest with large, mature birch. These varied habitats attract moose, white-tailed deer, black bears and coyotes. Wolves are often seen.

Mallards and other waterfowl nest near the harbor. Bald eagles, wood ducks and pileated woodpeckers may be seen during spring and summer, and in winter, gray jays and ruffed grouse are common. Great gray owls may be seen during some winters.

At least 16 kilometres of trails wind through the area, including a half-kilometre boardwalk along the sandy beach. Buffalo Bay also offers some of the best walleye fishing on Lake of the Woods.

Moose Lake and Birch Point

Moose Lake is at the western limit of the Great Lakes-St. Lawrence forest region. Eastern meadowlarks and black-throated blue warblers, near the limit of their breeding ranges, have occasionally been seen in the area. Southwest of the lake, watch for scarlet tanagers and indigo buntings. Between April and September, bald eagles, ospreys, whip-poor-wills, yellow-bellied sapsuckers and olive-sided flycatchers are all regularly seen. A walk along old logging roads may turn up a spruce or ruffed grouse.

Brian Wolitski

Though this tiny moose calf is still a little unsteady on its feet, by the time it's a few days old, it will be able to outrun a human, and swim readily.

Follow Moose Lake Road past the camping area for seven kilometres to reach Birch Point, a picnic area on Lake of the Woods. In the mixed woods along the road, brown creepers and black-throated green warblers may be seen. In May, the high, trilling song of the winter wren may be heard from brush piles and undergrowth in the forest. The lakeshore at Birch Point is a prime location for seeing northern parula, a tiny yellow-throated, blue-headed warbler with a buzzy, rising song, as well as great blue herons, American white pelicans, ospreys and common terns.

Access Buffalo Point is about 200 kilometres southeast of Winnipeg. Travel east on the Trans-Canada Highway (Number 1) to PTH 12 and then south to Sprague. Head northeast on PR 308 about 35 kilometres to the turnoff for Moose Lake and Birch Point, or continue for another 21 kilometres on PTH 12, then turn left on a gravel road to Buffalo Point First Nation. The year-round resort at Buffalo Point has a full range of facilities.

Fast Facts

Bordered by Manitoba, Ontario and Minnesota, Lake of the Woods is studded with an estimated 14,000 islands.

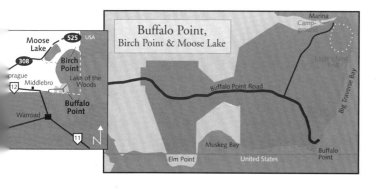

Eastern Manitoba

Whiteshell Provincial Park

Wildflowers create an ever-changing carpet among rugged granite outcroppings. Loons call across still lakes. Bears amble along forest paths of their own making. This rugged Canadian Shield beauty makes Whiteshell Provincial Park one of Manitoba's most popular, drawing active outdoors people, cottagers and visitors from far and wide.

Among Winnipeggers, the Whiteshell is popular as much for its accessibility as for its natural beauty. The 2,500-square-kilometre park, with its mighty rivers, ancient mountains and 236 clear, rock-edged lakes, is just an hour-and-a-half drive east of Winnipeg. The number and diversity of trails and canoe routes in the park mean there is always another path to hike or another lake to paddle. Snowmobile trails and 70 kilometres of groomed ski trails offer winter fun too.

The Trans-Canada Highway (Number 1) passes through the southern edge of the park, climbing gently from the Manitoba Lowlands into an expanse of Precambrian Shield bedrock some three billion years old. A meteorite crashing into the area long ago is believed to have formed the round, 110-metre-deep basin of West Hawk Lake, the deepest lake in the province. The Museum of Geological History, located next to the campground office, offers information about this crater, as well as Whiteshell geology.

The park offers other opportunities to walk back in time. Here, Anishinaabe ancestors created sacred petroforms centuries ago; later, fur traders paddled the swift rivers.

Although the southern end of the park is busy, particularly near the townsites of Falcon Lake and West Hawk, the Whiteshell is a frequent choice among backcountry hikers and paddlers. A 20-minute hike or paddle in along the waterways will find the

Archaeologists believe Manitoba's petroforms, such as this turtle, may be as old as 1,500 years. To help preserve and protect them, Manitoba's Heritage Resources Act makes it illegal to move or alter any of the stones.

Lichen and fungus cover the north side of this venerable aspen, opposite.

Even short trails in the Whiteshell can reward hikers with backcountry beauty.

crowds gone. Good road access and 325 kilometres of connected lakes, rivers and streams make the Whiteshell an excellent place for experienced and beginning paddlers alike.

The park has at least 15 hiking trails, as well as a number of biking trails. The rugged 66-kilometre Mantario Trail, highlighted later in this section, winds through the heart of the Whiteshell's wilderness. The South Whiteshell Trail links Falcon and West Hawk Lakes with a wide 12-kilometre track.

Fully serviced campgrounds and more than a dozen lodges and resorts can be found in the southern and western sections of the park. In the eastern Whiteshell, there is a large undeveloped Wilderness Zone.

Boreal forest covers 40 per cent of Manitoba and the Whiteshell is a good place to explore this lush and beautiful environment. Forested areas are typically made up of white spruce, Manitoba's provincial tree, and balsam fir, intermixed with trembling aspen, birch and balsam poplar. In poorly drained fens and bogs, tamarack or black spruce can be found. The hardy jack pine, with its scaly gray bark and paired needles, is one of the most common evergreens in the Whiteshell.

Fruit-bearing shrubs include saskatoons, chokecherries, wild plums and blueberries in abundance. In the fall, smooth sumac covers the ridges with brilliant red leaves. On the forest floor are bluebead lily, false Solomon's seal, nodding trillium, lily-of-the-valley and sarsaparilla.

Birds, black bears, turtles and beavers are the wildlife to look for here. Owls, including Manitoba's provincial bird, the great gray owl, may be seen in any season, especially in secluded tamarack bogs. In summer, ospreys swoop down on fish swimming near the surface and ruby-throated hummingbirds flit from flower to flower. In September and October, many of the shallower lakes offer resting spots for migratory waterfowl, including common mergansers, common goldeneye, bufflehead and lesser scaup. In November, when mists form over the freezing lakes and the migratory birds depart, evening grosbeaks, chickadees, blue and gray jays and spruce and ruffed grouse can be readily seen in the forest. Bald eagles and hawks soar overhead or perch on the taller trees in the area.

Turkey vultures, often mistaken for eagles or ospreys, are seen sailing high in the skies from spring to early fall. Seldom flapping their wings, they rock from side to side in a distinctive flight pattern.

Black bears are plentiful in the Whiteshell and campers, hikers and motorists often encounter them. Motorists on the Trans-Canada Highway should be alert for white-tailed deer, especially in the early morning and just before dusk, feeding along the verges. Red foxes can also be seen along the roadsides. Paddling quiet streams and rivers may reveal a moose at the water's edge or snapping turtles sunning themselves on logs and rocks. Keep a sharp eye out for mink, beavers and river otters. There are also lynx in the park, but they are rarely seen.

Barbara Endres

Omnivorous and agile, black bears are at home in trees or water, or on the ground, where they have been clocked at speeds of up to 55 kilometres per hour. Like humans, bears plant their entire foot on the ground with each step, in a gait that is slightly pigeon-toed. The feet are well furred and have long claws, which are often used for tearing apart old stumps or logs in search of food.

Access West Hawk Lake, Falcon Lake, Rennie, Seven Sisters Falls, Pinawa and various resorts or campgrounds are good staging areas in which to stay overnight, buy supplies or meet companions before heading deeper into the Whiteshell in pursuit of wildlife. There are four main road entrances to Whiteshell Provincial Park. The Trans-Canada Highway (Number 1) to Falcon and West Hawk Lakes in the southern portion of the park is most popular, but access to the central part is best on PTH 44, while PR 307 leads to the northern Whiteshell. This is the route to take to the Pine Point Trail and the Bannock Point Petroforms. Access to the Mantario Trail (north trailhead) and many canoe routes is via PR 309, which branches east off 307. Access to the northernmost part of the park along the Winnipeg River is via PR 313 to Pointe du Bois.

Eastern Manitoba

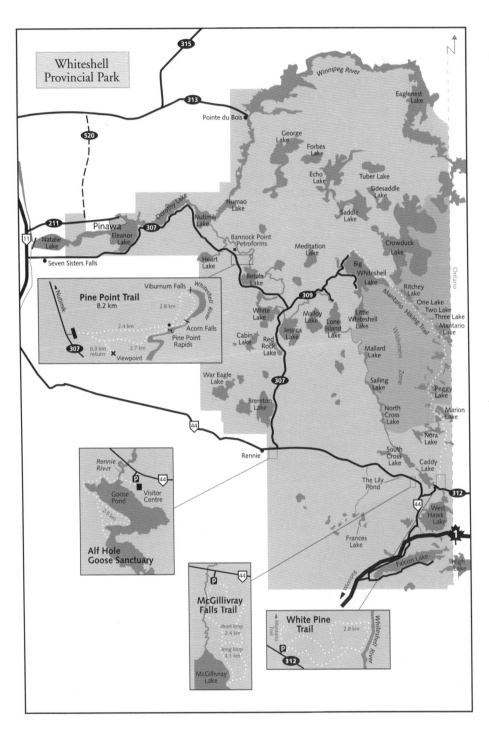

Alf Hole Goose Sanctuary

The sanctuary was named after a mink rancher, who more than 60 years ago bet a crock of whisky he could keep four abandoned goslings alive for six weeks. Now, the self-guiding trail, observation gallery and interpretive centre make it one of the better places in the province to view waterfowl at close range.

More than 150 Canada geese summer here every year. Many arrive here after making a 1,000-kilometre flight, often non-stop, from their winter refuge in southeast Wisconsin. Hundreds more stop by during May and October migrations.

Goslings hatch from mid-May to June. In June, visitors can also expect to see nesting pairs of wood ducks as well as several other species of dabbling ducks, including green-winged and blue-winged teal, American wigeon and, rarely, American black ducks. At least one pair of common loons resides on the lake and red-necked and pied-billed grebes are often seen. Eastern wood-pewees nest nearby; their plaintive song can be heard in June: *pee-oo-wee, pee-oo.*

Visitors in late June and early July may see moose with their spindly legged calves. There are several beaver lodges on the pond and mink frogs can be found here. Snapping and western painted turtles are sometimes found basking on rocks or logs.

Photographer's Notes

June 8th: "An incredible viewing area for geese and wood ducks; one of the best in the region. Park interpreters are helpful and knowledgeable. The place is a real gem."

Among the Canada geese that spend each summer at the sanctuary, between 10 and 12 pairs breed and raise a family like this one.

Brian Wolitski

Access: The sanctuary, with its interpretive centre, is one kilometre east of Rennie on PTH 44.

Eastern Manitoba

Lily Pond

The rock that cradles the Lily Pond is part of the primeval Precambrian Shield, created more than three billion years ago as the Earth cooled. The basin that holds the water is much younger; it was sculpted by advancing glaciers some 20,000 years ago as the Laurentide ice sheet moved south over the ancient bedrock.

The rock soars in high cliffs surrounding the pond. On weekends, visitors will often see climbers on the rock faces.

This is one of the most accessible places in Manitoba to see and smell sweet-scented white water lilies, with peak blooming time in July. The pond also supports rare mink frogs, which can be seen on or around the lily pads beginning in early June. Early in the morning, their *chuk chuk* sounds – like nails being driven in the distance – can be heard.

The wetland also attracts muskrats, and mallards arrive in April to breed.

 A sign along PTH 44, just west of West Hawk Lake, marks the Lily Pond.

Mink frogs, right, and white water lilies combine to make this a special place.

Snapping turtles have powerful jaws that can slice through a thumb-sized branch.

McGillivray Falls Trail

This 2.8-kilometre or 4.6-kilometre self-guiding loop provides interesting lessons about the drainage of the Whiteshell, a system that is termed "deranged" because the water follows glacier-cut channels. Here, rock surfaces and steep inclines lead to McGillivray Lake. Its drainage basin offers a small example of the movement of water in the Whiteshell, Rennie and mighty Winnipeg River systems that is typical of the Precambrian Shield. Black spruce and bog plants grow on the lower side; aspen, birch and white spruce blanket the hills on the higher side. Hardy jack pines cover the ridges.

Water cascades over granite at McGillivray Falls, imperceptibly wearing away the ancient rock.

Many species of wildflowers can be seen along the trail, including stemless lady's slippers, also called moccasin flowers. Like many shield lakes, McGillivray is shallow, at most only three metres. Naturally occurring humic acid gives the water a brownish tealike tinge. Bald eagles and ospreys may be seen patrolling the lake for fish.

Evidence of the tireless activity of beavers is everywhere; in many places their dams have turned the stream into small reservoirs, slowing the force of the water. This is also prime moose country and hikers in the early or evening hours might spot an animal feeding in the open meadows. White-tailed deer are also often seen.

In late May and June, common mergansers – larger and less flamboyantly crested than their hooded or red-breasted cousins – and common loons can be seen courting on the lake. Alder flycatchers build nests along the shore.

Access: The turnoff to the trailhead is on PTH 44, 2.7 kilometres north of the junction with PR 312.

Eastern Manitoba

White Pine Trail

This 2.8-kilometre trail winds over rock formations to the Whiteshell River and then loops back to its starting point through a wooded area marking the transition between the boreal and Great Lakes-St. Lawrence forest habitats. Eastern white pine, rare in most of Manitoba, can be seen.

The first part of the circuit passes through a forest of broken trees, mostly jack pine, felled during a severe windstorm in 1982. A white spruce more than 90 years old survived; it's too large to encircle with your arms. On the opposite bank of the river is the Whiteshell Fish Hatchery Interpretive Centre. Along the river eastern white cedar can be found, along with black ash, balsam fir and the magnificent eastern white pine. In the shade of the forest the stemless lady's slipper can be found, its deep pink blossoms glowing among the shadows.

The old trees are a magnet for woodpeckers. Pileated, downy, hairy, three-toed and black-backed woodpeckers may be seen here. Look for large holes in the base of tree trunks, evidence of the large pileated woodpeckers.

In early to mid-May, this trail can be an excellent location to observe warblers during spring migration. Common goldeneye are often heard before they are seen; their whistling wings herald their approach as they fly along the course of the river. The forest floor is carpeted with bunchberry, its white flowers blooming in June and the berrylike red fruit appearing in August. The fruit attracts white-tailed deer and ruffed grouse. Later in the season, spruce grouse are drawn to common juniper by its aromatic berrylike cones. Particularly in winter, deer eat the branches as well as the berries.

Though this trail was created to celebrate the majestic white pine, which is rare in Manitoba, its hardy, shallow-rooted cousin, the jack pine, below, is most prevalent here.

Ian Ward

Access The trailhead is on the north side of PR 312, north of West Hawk Lake. The Whiteshell Fish Hatchery Interpretive Centre is east of the White Pine trailhead on PR 312.

Pine Point Trail

North of Betula Lake, the 10-kilometre Pine Point Trail winds through mixed forest and bedrock covered in multicolored lichens. The first five-kilometre loop leads through boreal forest to a beautiful picnic spot overlooking Pine Point Rapids on the Whiteshell River. From there, a second loop continues to Acorn and Viburnum Falls. Here, in spring and summer, look for spotted sandpipers hopping from boulder to boulder in search of food.

At times wide and fast, at times shallow and filled with wild rice, the Whiteshell River winds northward through the park from West Hawk Lake to Nutimik Lake. It is one of the first areas of open water for returning waterfowl in the spring. In May, Canada geese, mergansers, common goldeneye, ring-necked ducks and lesser scaup may be seen. The surrounding forests are alive with birds and blossoms as highbush cranberry, downy arrowwood and nannyberry are all in flower. In June, with luck, one may see the moccasin flower or stemless lady's slipper, a large and beautiful pink orchid. Taking advantage of the bounty, swallowtail butterflies can be found in abundance.

On sunny days in spring and summer, snapping turtles can be seen basking on logs or rocks. By August, red squirrels are beginning to cache seeds and nuts for the winter. In September, migrating waterfowl return to glean wild rice along the shallows where the river widens, sharing the harvest with aboriginal Manitobans, who have gathered wild rice from this river system for centuries.

In the winter months, skiers may spot signs of moose, white-tailed deer, snowshoe hares, wolves or lynx.

 The trailhead is southeast of Nutimik Lake off PR 307.

Bannock Point Petroforms

These petroforms – stones laid out on the bedrock in the shapes of snakes, turtles and birds – are believed to have been created about 1,500 years ago by aboriginal Manitobans.

Today's Anishinaabe continue to use the sites, which are located northwest of Betula Lake, treating them with care and respect, as all visitors should.

Moose, like this big bull, above left, can be found where the river widens into shallows, or in forest clearings where sunlight encourages tender undergrowth.

Mantario Hiking Trail

This 66-kilometre trail is the longest hiking trail in Western Canada's shield country. Because of its remote location, the possibilities for viewing wildlife are excellent. This is one of Manitoba's best-known and most arduous trails, created to provide exceptional terrain for experienced backpackers.

Passing through stands of jack pine along high rock ridges, winding through forests of white spruce and often descending into creek beds and bogs filled with black spruce and tamarack, the trail is always varied. In addition to spruce and pine, hikers will see aspen and balsam fir, mountain maples, saskatoons, blueberries, chokecherries, wild plums and a sweep of wildflowers and lichens. In the spring, masses of blossoms attract butterflies, while the resulting fruits attract bears, grouse and squirrels in summer and fall.

Those attempting the trek may be faced with challenging beaver dams, wet, slippery rocks, and climbs up some of the highest rock ridges south of the Winnipeg River. But there are bonuses, for hikers may be treated to the sounds of howling wolves, the signs or sight of a moose in a bog or a black bear on a ridge, and a continuous acquaintance with the numerous birds of the boreal forest. Bald eagles, turkey vultures, spruce and ruffed grouse and blue and gray jays are common throughout the area.

The Mantario Wilderness Zone is popular with hikers and paddlers in summer and cross-country skiers in winter. Wolf tracks are often seen on frozen lakes as well as along forested trails.

The three-to-six-day Mantario Trail is for experienced backpackers, in good shape and familiar with wilderness camping. In places the trail can be overgrown and difficult to follow. A shuttle must be arranged for those wishing to do the entire trail. Late summer, from the end of August to the end of September, is the best time to go. Leaves turn, mosquitoes and black flies disappear and temperatures cool to provide comfortable and pleasant hiking. May and June may be wetter and rainfall greatly affects trail conditions. July and August can be hot and though the trail is popular in the summer, biting insects are often a nuisance.

Hikers are almost certain to see bald eagles, a fitting symbol of unspoiled wilderness, along the Mantario Trail.

The trail may be accessed from south or north. From PTH 44, head east 3.5 kilometres on PR 312 to the southern trailhead, east of Caddy Lake on the north side of the road. To reach the northern trailhead at the north end of Big Whiteshell Lake, enter the park via Rennie on PTH 44. Head north on PR 307, then east on PR 309 to the Mantario Trail sign. A trail brochure with a 1:50,000 contour map is available from the Department of Natural Resources.

The Manitoba Naturalists Society offers a guided five-day canoe trip to a cabin on Mantario Lake in the park's Wilderness Zone. The canoe route roughly parallels the Mantario Trail as far as Mantario Lake.

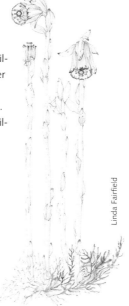

Found in deep, moist shade, often among mature stands of aspen, the stems of Indian-pipe, left, are a nearly translucent white, while the waxy flowers may have a slightly pinkish tinge.

Seven Sisters Falls

Devoted parents who add to their substantial nests each year, ospreys live exclusively on fish.

According to archaeologists, Seven Sisters Falls has been considered a special place for more than 6,000 years. Recent excavations show the land was used as both a communal meeting place and a native burial ground.

In our time, this area – where aspen parkland melts into thick boreal forest – is a favorite destination for avid birders, fishing enthusiasts, canoeists and those looking for quiet charm.

There are two major attractions here – the Seven Sisters Hydro Dam and Whitemouth Falls Wayside Park. The dam, on the Winnipeg River, created a huge reservoir to the east. Called Natalie Lake, it has become a year-round hot spot for birders.

The picnic site at Whitemouth Falls is cradled in a stand of oak trees, sprinkled with white spruce, green and black ash, birch and aspen. Raspberry, blueberry, pincherry, chokecherry, cranberry, gooseberry and wild plum shrubs can be found, along with abundant roses. The area is carpeted with bluebells and irises, which also grow among the rocks leading down to the rushing falls. A walk west leads to dark spruce bogs ringed with black spruce and tamarack.

As in much of Manitoba, white-tailed deer abound here. There are black bears in the area, especially during July and August – berry-picking time – along with red foxes and wolves. The last of these are often heard but rarely seen. Otters frequent the shoreline, and farther along the river, in quieter spots, look for beaver activity and muskrats.

Among birders, the Seven Sisters area is considered to be one of the best boreal forest birding places in the province. In May, it is a wonderful place to see yellow, bay-breasted, Cape May and chestnut-sided warblers, as well as American redstarts. Other birds that are easily seen include bald eagles, ospreys and American white pelicans, which have been regular visitors to the

Gulls and terns and, in recent years, American white pelicans, are among many species that find Whitemouth Falls a bountiful place.

falls for the past decade. Double-crested cormorants, common goldeneye, hooded and common mergansers, and Caspian and common terns can be sighted during spring and fall migration.

At Natalie Lake, birders armed with binoculars will be rewarded with numerous sightings from the walkway atop the hydro dam. A closer look can be had by hiking along the dikes, which stretch along the north and south sides of the lake. In spring (early May to mid-June) and fall (mid-September to late October), the lake attracts common loons, horned and red-necked grebes, greater and lesser scaup and bufflehead. Oldsquaw and all three scoters – black, white-winged and surf – have been sighted here.

Hundreds of cliff swallows nest on buildings at the dam; among them, rough-winged swallows, with their dull brown breasts and backs, may be seen. Great blue herons often wade along the shallows of the Winnipeg River. In some winters, PR 307 and other roads in the area offer an excellent chance of seeing great gray owls.

Access The town of Seven Sisters Falls is located about 75 kilometres northeast of Winnipeg. Follow PTH 59 north to PTH 44 and head east to the junction of PTH 11. Here, turn north on PTH 11 for five kilometres and then go east on PR 307. The dam and Whitemouth Falls Wayside Park are on the north side of the road, just before you reach the townsite.

Pinawa

Pinawa's parklike setting, waterways and surrounding boreal forest combine to make this small eastern Manitoba town an excellent birding and wildlife viewing area. Located on the winding Winnipeg River, which makes its way from Lake of the Woods in Ontario to Lake Winnipeg, the town and its environs attract more than 100 bird species.

The Ironwood Trail can be accessed at Aberdeen Avenue and Willis Drive in Pinawa and is named for the rare hop-hornbeam, or "ironwood" trees found along the trail. The trail follows the Winnipeg River where visitors will often see common goldeneye, as well as American white pelicans and double-crested cormorants in migration. In May, yellow-rumped and Nashville warblers arrive, as do American redstarts and Baltimore orioles. In August, the area's many oak trees are alive with some 20 species of warblers, including bay-breasted warblers and northern parula, and four species of vireos. In September, Harris's sparrows feed along the trail and sandhill cranes circle in migratory flocks overhead. In winter, one can often see the yellow, red and blue of evening grosbeaks, pine grosbeaks and blue jays adorning the feeders in town.

At the end of PR 211 and across an old dam, an excellent network of cross-country ski trails turn into hiking trails in summer, though some may be wet or flooded. From mid-May to early July,

Hikers along the Pinawa Channel, right, are likely to see a variety of bird species at almost any time of the year.

Common goldeneye, such as this handsome male, return every April to raise their broods along the Winnipeg River.

expect to see many warblers, including mourning and Connecticut warblers. In fall and winter, hikers and skiers may see three-toed and black-backed woodpeckers, as well as spruce grouse.

About two kilometres west of Pinawa is Pinawa Cemetery Road. Big white-tailed deer bucks frequent the area around the intersection. Park across from the cemetery and hike or bike north a few kilometres along the road. In late May, some 20 species of wood warblers can be found here, including Canada and black-throated green warblers. Listen for the *please-please-ta-meet-you* song of the chestnut-sided warbler.

An unmarked hiking trail leads north to a wetland on the Pinawa Channel, a branch of the Winnipeg River. Better yet, paddle five or six kilometres along the channel, starting at the cemetery and taking out at Pinawa Dam Provincial Heritage Park. Expect to see black terns, swamp sparrows and bald eagles.

Mammals such as black bears, wolves, beavers and moose also live here and tracks can often be seen throughout the Pinawa area. More than half of Manitoba's 150 butterfly species can also be found.

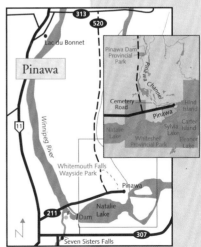

Access From Winnipeg, take PTH 59 north to PTH 44. Travel east on PTH 44 until you reach PTH 11. Head north on PTH 11 for 10 kilometres to the turnoff on the right for PTH 211 and Pinawa, about 10 kilometres east. A drive up PR 520 north of Pinawa offers good birding and possibilities of seeing deer, foxes, bears, coyotes or, if you're lucky, wolves.

Eastern Manitoba

Nopiming Provincial Park

Nopiming Provincial Park is a vast area of Precambrian rock, boreal forests and clear lakes, just two hours from Winnipeg. In Anishinaabe, the name means "entrance to the wilderness" and though mining and logging occurs in Nopiming, and cottages can be found in its southern sector, the park nevertheless provides prime wildlife viewing opportunities.

A common loon on a sparkling lake, above, or a black bear, like this magnificent nibbler, left, may both be seen in Nopiming.

The story of Nopiming goes back billions of years. This is shield country, a place where the primeval crust of the earth was thrust into towering mountains more than 2.5 billion years ago. The resulting rocks are extraordinary, their ancient origins evident in colors of gray and pink, streaked with black and white, sometimes in patterns resembling waves.

In the vast stretch of time since, wind and water eroded those great, jagged peaks into low, undulating ridges; the landscape received its final "touch-ups" a mere 20,000 years ago when enormous glaciers scoured the land and left behind deposits of clay and gravel as they retreated. From afar the rocks appear to be covered in great splashes of green and black paint; a closer look reveals lichen, delicate and brittle, thriving where little else can grow.

People have been living in Nopiming for at least 9,000 years, according to archaeological excavations at Caribou Lake. Digs in other locations have uncovered pottery, as well as cutting and scraping tools and bones from later cultures.

The forests of Nopiming are thick with jack pine, trembling aspen, white spruce and balsam fir; bogs are scattered throughout the park – easy to locate if you look for typical stands of black spruce and tamarack. Hazelnut, alder and mountain maple fill in the underbrush.

Though only thin soil covers the rocky base, the landscape is nevertheless filled with colorful plants and beautiful wildflowers. Wild sarsaparilla is followed by yellow evening primroses, one-flowered wintergreen and bunchberry. You can find raspberry, three-toothed cinquefoil, narrow-leaved meadowsweet, white

Roger Turenne

All species of the bat family known to occur in Manitoba can be found in Nopiming – ubiquitous little brown bats, like the one above, rare northern long-eared bats, as well as hoary, silver-haired, big brown, and red bats. All live exclusively on insects.

cinquefoil, wild strawberry and yellow avens. Equally abundant are Manitoba native grapes, fireweed, Canada anemone, nodding trillium and lovely wild columbine.

Though Nopiming boasts a varied and abundant wildlife population, it's the woodland caribou and prolific colonies of bats that give the park a special quality. This is the home of the most southerly herd of woodland caribou in Manitoba. Called the Owl Lake herd, it numbers about 50 animals, which generally live in and around Owl and Flintstone Lakes. More information about the herd can be found at an outdoor exhibit at Black Lake campground.

For the "bat show", head to Tulabi Falls campground at dusk, for these shy little mammals prefer to sleep all day (upside down) and feed at night. Nopiming is also one of the best parks in the province in which to see pine martens. This athletic member of the weasel family can be found in trees along lake trails, leaping and running after squirrels. Other furbearing mammals include playful otters, mink, beavers, muskrats, fishers, short-tailed weasels, porcupines and a very few wolverines. Nopiming also boasts wolves, coyotes, red foxes, lynx and black bears.

The wetlands and large areas of new forest growth that followed a series of fires in the 1980s provide plenty of delectable green shoots and tender twigs for moose and white-tailed deer.

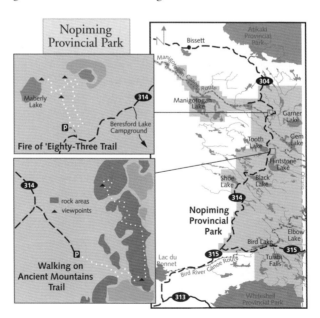

Common goldeneye ducklings await the return of their parents, a picture of sibling bliss.

The new growth also supports a huge diversity of small mammals: woodchucks, chipmunks, shrews, including masked, water, Arctic and pygmy shrews and voles, including heather, western red-backed and meadow voles. Northern bog lemmings and several species of mice, including the meadow jumping mouse are common. Red and northern flying squirrels and snowshoe hares also flourish here.

Amphibians and reptiles thrive along the lakes and rivers; a careful wait might reveal a blue-spotted salamander or a mudpuppy. There is a whole range of frogs to watch and listen for – eastern gray tree frogs, boreal chorus, wood and leopard frogs, and possibly green and mink frogs.

Snapping turtles bask along the shores and on rocks, as do large painted turtles, while slithering through the park are a variety of Manitoba's harmless snakes such as northern redbelly snakes, red-sided garter snakes and smooth green snakes.

There are an estimated 188 species of birds in Nopiming, including hawks, sandpipers, owls, woodpeckers, flycatchers, swallows, chickadees, nuthatches, wrens, warblers, blackbirds, finches, gulls, sparrows and herons. Lesser scaup, blue-winged teal, buffleheads, ring-necked ducks and common loons may also be seen during spring and summer. Wild rice, which grows in many of the small lakes, provides a harvest for humans, and also attracts waterfowl, including wood ducks and mallards in autumn.

Bald eagles and turkey vultures soar above the trees; below, the forests support an abundance of ruffed and spruce grouse.

Nopiming is located about two hours northeast of Winnipeg via PTH 59 and PR 317, 313 and 315.

Photographer's Notes

June 9th: "In one day in Nopiming we have seen deer, three bears, bald eagles, hummingbirds, a red fox, moose sign, snapping turtles and a host of forest and water birds. Wildflowers everywhere."

Eastern Manitoba

Fire of 'Eighty-Three Trail

This trail allows visitors an opportunity to look at fire, not as a destructive force but as a life-renewing power in Canada's vast northern forests. Stretching from Alaska to Newfoundland, the boreal forest ecosystem has existed for at least 8,000 years. During those eight millennia, wild fire has kept it in a constant state of change and revitalization.

One of the largest blazes in Nopiming was the Long Lake Fire, which began in early September 1983. The trail was created to interpret the enormous change it caused. Following a self-guiding brochure, visitors can observe a forest in renewal.

The trail is a short, fairly demanding hike over an uneven rock surface, following a loop with a small spur leading to Maberly Lake. The path winds through thick stands of young jack pine, a tree that might be called the Sleeping Beauty of the boreal forest. Jack pine cones lie patiently in the forest, year after year, locked tight by hard resin, waiting for the searing heat of a forest fire to burst them open and scatter their seeds on the warm ground, where they germinate and take root.

The young aspen along the trail also owe their lives to fire. These small trees sprang up from parent roots that remained alive under the ground. However, as post-burn mates, aspen and jack pine compete for sunlight, moisture and nutrients. The jack pines seem to be winning the battle here.

Farther along the trail, snags – tall, charred trees – serve as forlorn reminders of the destructive qualities of a forest fire. But in nature, dead tree trunks provide specialized habitat for a multitude of animals. Insects like bark beetles and flathead borers thrive

This three-toed woodpecker is just one of many species that benefit from periodic wild fires.

on dead trees, attracting, in turn, a variety of woodpeckers. Tall snags provide ideal nesting and perching places for ospreys, bald eagles, northern hawk owls, swallows and flycatchers.

There are many "fire survivors" along the trail. It's commonly thought that fire destroys everything in its path; in fact, most fires consume only a third of the living material in a burn area. Because forest fires have been frequent in Nopiming, it's possible here to examine young growth alongside older growth, which in turn grows near even older growth.

From late June through October, wild strawberries, raspberries, blueberries, saskatoons, pincherries, chokecherries, rosehips and bearberries ripen one after another. These are favorite foods for many species, including a variety of songbirds, spruce and ruffed grouse, black bears and coyotes.

Halfway along the trail a half-kilometre long spur trail leads to Maberly Lake. The extra hike offers an opportunity to see an unburned section of mature forest. Around the lake an old growth forest of black spruce, tamarack and scattered balsam fir supports pileated woodpeckers and the elusive lynx, which thrives on the snowshoe hares, grouse and red squirrels living in the young forest nearby.

There are generally one or two common loon families on the lake. The mating displays of the males can be seen in late May and early June; in July and early August, the young can often be seen. In fall, Maberly Lake is a resting place for Canada geese and other migrating waterfowl. You may also see a moose along its shores.

Their seeds released by fire, jack pine seedlings soon cover the forest floor after a major blaze, as they have here in Nopiming.

 The Fire of 'Eighty-Three Trail is at the northern end of the park, just west of PR 314.

Walking on Ancient Mountains Trail

This rigorous 1.8-kilometre trail introduces visitors to the magnificent three-billion-year-old rock of the Canadian Shield. Particularly in the spring, the natural beauty of some of the oldest rock in the world dramatically blends with the delicate freshness of new forest growth. Toss in a panorama of splendid views and it's little wonder this trail will become an unforgettable memory.

It's because of a recent natural event that visitors have the chance to observe the geological past so clearly. In early September 1983, after a prolonged drought, lightning sparked a forest fire that burned an area of 25,420 hectares. When the ashes cooled the forest vegetation was gone; in its place, the park's foundation was exposed.

Dennis Fast

Though this youngster is a picture of defiance, northern hawk-owls are quite tame and, unlike many owls, hunt during the day.

Rock is the predominant feature, but the landscape is anything but lifeless. Here, you can see the variety, color and life of rocks, created by heat and pressure, sculpted by the icy fingers of moving glaciers. Following the excellent self-guiding trail brochure, you'll first be introduced to sedimentary rock – light gray and brown layers that 2.7 billion years ago were flat beds of sand and mud spread along a great sea floor. Over the millennia, this seabed was pushed deep into the earth, where heat and pressure transformed it to hard rock.

Farther along is a white rock called pegmatite, sprinkled with crystals of grey quartz, pale pink feldspar, brown mica and black tourmaline. Picture it billions of years ago, a molten liquid squeezing up between cooler sedimentary rocks before solidifying.

Among the ancient rock formations, islands of young jack pine serve as living reminders of the 1983 fire. One of the first trees to appear after a burn, jack pine is a fine example of nature's adaptations, for it requires the intense heat of a fire to reproduce. The searing heat opens the cones and allows the seeds to germinate.

As you walk the trail, the difference between this new forest and an old growth forest is obvious. Old forests are hidden places, shadowed by thick growth. This young forest glistens with a green brilliance made even more striking by lush stands of purple fireweed.

The summit of the trail is a huge rock outcrop with a spectacular 360-degree view. From this spot the trail circles the peak, providing magnificent vistas in all directions. When you get back to the summit, retrace your steps to the trailhead.

 This trail offers opportunities to see or hear a variety of Nopiming's birds. Red-breasted nuthatches, and occasionally northern saw-whet owls, are among the park's year-round avian residents. Also listen for the slow jackhammer rhythm of pileated woodpeckers and the softer, swifter tapping of black-backed woodpeckers as they hunt beetles in the charred tree trunks. The croaking call of ravens on the wing can often be heard and northern hawk-owls may be seen, perched atop dead snags.

Several species of warblers nest in Nopiming. In June, Blackburnian warblers, distinguished by their high, thin songs and bright orange throat of the male, can sometimes be found in the tops of the young pines. Bay-breasted and blackpoll warblers can also be seen, particularly during spring migration in May.

Dead tree stumps and fallen trunks attract beetles and ants and these in turn provide food for black bears. Early morning and dusk are the best times to see moose and white-tailed deer. Porcupines are here too, as are woodchucks, chipmunks, squirrels and voles. Snowshoe hares are sometimes seen; their presence attracts red foxes, wolves, coyotes and lynx, but the chance of seeing any of these elusive hunters depends on patience and luck.

Access: The trailhead is east of PR 314 in central Nopiming Park, just east of Tooth Lake. Hikers are advised to wear footwear that grips, especially during wet weather. Small children may find the trail difficult.

Unveiled by fire, Nopiming's ancient mountains provide spectacular panoramas, as well as an opportunity to examine the rock that serves as the foundation of much of Canada.

Jerry Kautz

83

Bird River Canoe Route

For most canoeists, this route begins at the Tulabi Falls campground. Before putting in, however, the campground itself is worth a visit. One of the more spectacular sights here is Tulabi Falls, with a viewing platform that brings you within a few feet of the tumbling water. Pink wintergreen, twining honeysuckles, nannyberries, rattlesnake ferns and three-toothed cinquefoil can all be seen along the paths around the falls. And as indicated in the park overview, this is one of the best places to view bats in Manitoba.

The route begins with an eight-kilometre paddle along Bird Lake, with its cottage-lined shores, and an equally placid nine kilometres toward and beyond PR 315 (an alternate starting point). The river becomes more challenging as Bernic Lake Road comes into view, with rapids, log jams and the first of a series of portages.

The balance of the route is varied and often demanding, with smooth stretches alternating with whitewater, falls, a canyon and portages before the last set of easy rapids just upstream of the pullout at the Pioneer Beach Campground.

It appears that regular traffic along the Bird River has habituated the resident wildlife to human activity and many species seem quite comfortable in the presence of paddlers.

Along this route, be on the lookout for moose, white-tailed deer, black bears, muskrats, martens, otters, beavers and weasels. An abundance of birds includes ospreys, herons, kingfishers and bald eagles. The great gray owl is also regularly seen in this area.

Though it runs through Nopiming's most populous region, portions of the Bird River, such as this section between Tulabi Falls and Bird Lake, aptly reflect the meaning of the park's name – "entrance to the wilderness".

Manigotagan Canoe Route

Following the Manigotagan River from east to west, the route links Long, Manigotagan and Quesnel Lakes, and continues on to Lake Winnipeg. There are access points at PR 314 and at Caribou Landing on Quesnel Lake. If you put in at PR 314, your stamina and endurance will be quickly tested on a series of rapids about 1.5 kilometres downstream, with subsequent portages through thick deadfall – reminders of the great fire of 1983. The next stretch of the river is peaceful, winding through low, swampy country to Long Lake, with its shoreline cottages and motorized watercraft. On leaving Long Lake, white and black spruce, balsam fir and trembling aspen line the shores and the river tumbles westward in a series of waterfalls en route to Manigotagan and Quesnel Lakes. Quesnel Lake is a popular departure point for canoeists who prefer the shorter 76-kilometre, three-day route to the community of Manigotagan.

Dozens of species, including this busy muskrat, share the Manigotagan River with paddlers.

From this point, the route is clearly divided into three manageable sections: the first, to Pillow Falls, has many waterfalls and portages; the second section, from Pillow Falls to Skunk Rapids is gentler, with easier rapids, and the final stretch, from Skunk Rapids to Manigotagan, is particularly challenging, with difficult rapids and rocks.

The diverse habitats along the river support many of Nopiming's small mammals, and a number of its large ones. In particular, look for moose and occasionally, in the spring, woodland caribou along the river between Long and Manigotagan Lakes. Black bears are often seen ambling along the shoreline or swimming the calmer reaches of the river. Fortunate paddlers may share their river journey with otters that swim and play alongside their canoes.

Bald eagles, ospreys, American white pelicans, red-tailed hawks, herring and ring-billed gulls, terns and even turkey vultures are common sights along the river, as are western painted and common snapping turtles, frequently seen sunning on logs or rocks along the shore.

Eastern Manitoba

Atikaki Provincial Wilderness Park

Situated along the Ontario border, Atikaki Provincial Wilderness Park lies north of both Nopiming and Whiteshell Provincial Parks. Atikaki means "country of the caribou" and the park is a place of bountiful wildlife, beautiful scenery and challenging canoe routes.

Accessible only by air or by water from Wallace Lake and PR 304, this park encompasses almost 4,000 square kilometres of rock outcrops and granite cliffs interspersed with a complex of bogs, fens, marshes, rivers and forest. Three rivers, including the Manitoba portion of the Bloodvein River, a Canadian Heritage River and Atikaki's crown jewel, flow westward through the park. The "pool-and-drop" Bloodvein originates near Red Lake, Ontario, and tumbles over a series of granite ledges and through placid lakes until finally emptying into Lake Winnipeg 306 kilometres west.

The Leyond River joins the Bloodvein near Lake Winnipeg and the Pigeon River plunges through canyons farther north, providing whitewater rafting or paddling that rivals the most celebrated rivers in the world.

Early artists inscribed many of Atikaki's walls of rock with spectacular pictographs depicting bison, human figures, hands and power symbols. Undisturbed archaeological sites along the Bloodvein corridor provide strong evidence that people have lived here for more than 6,000 years.

Sedge meadows and lowland balsam fir and balsam poplar forests, carpeted with lichen, provide prime woodland caribou habitat. Floating bogs of sphagnum are dotted with Labrador tea, bog laurel and leatherleaf, and fringed with alder and willows. Wild rice grows along the shores of slow-moving rivers and in the bays of shallow lakes.

Its remote nature allows opportunities for viewing black bears, moose, woodland caribou and a host of fur-bearing animals. Moving quietly along even a small section of the 600

Happiest when knee-deep in quiet water, moose are also at home on land. When treading muskeg or soft shorelines, their cloven hooves (as much as 15 centimetres long) spread widely to support the animal's great weight.

Eastern Manitoba

Two types of lichen – light-colored reindeer lichen, inset, and oldman's beard, which hangs from branches of mature trees – are important foods in the diet of woodland caribou, above.

kilometres of interconnected waterways in a canoe is a good way to see wildlife here. The rivers and lakes are havens for beavers, river otters and muskrats, as well as painted and snapping turtles. The water teems with walleye, northern pike and lake trout.

Manitoba has a resident population of woodland caribou estimated at between 2,000 and 2,500 animals. The Committee on the Status of Endangered Wildlife in Canada has designated these secretive animals as vulnerable west of Ontario. By comparison, the largest barren-ground caribou herd that visits the province's northern reaches during the winter is 200 times larger, numbering nearly a half-million animals. Unlike their barren-ground cousins, woodland caribou are sedentary, living in small, scattered groups that rarely move more than a few kilometres.

Sasaginnigak Lake in Atikaki Park is a large, island-studded boreal lake often used by woodland caribou during the calving season in spring. Paddling on this lake may afford the sight of a half-dozen caribou swimming from one island to the next, seeking refuge from wolves.

Dependent on lichens as a food source, they are particularly vulnerable to logging or fires, which replace old-growth forest and lichens with deciduous trees and shrubs. This environment attracts competing populations of moose and white-tailed deer, which in turn support an increased number of predatory wolves. Woodland caribou seldom survive the loss of the forests in their home territory.

Though rarely seen, fishers, medium-sized carnivores with thick, dark fur, also live in the forests; wolves are less secretive and can often be heard at night. A wolf howl at dusk may elicit a response. Bald eagles and ospreys soar overhead throughout the summer.

Access While there is no direct road access into the park, more than a half-dozen lodges, including one on Sasaginnigak Lake, offer fly-in access and accommodation. The most popular water route begins at Wallace Lake, south of the park. From Winnipeg take PTH 59 north for about 125 kilometres to PR 304. Continue past Bissett to Wallace Lake. The Bloodvein River may be accessed from several places along its route, but one popular route starts from Aikens Lake down the Gammon River to its confluence with the Bloodvein, then continues northwest to where the Leyond River joins it, and ends at the Bloodvein First Nation on the shores of Lake Winnipeg, a total of 200 kilometres. Numerous tour operators run trips on the Bloodvein from Winnipeg. Contact Travel Manitoba or a travel consultant for a list of contacts. Experienced paddlers can plan their own trips with maps and air photos available from Manitoba Natural Resources.

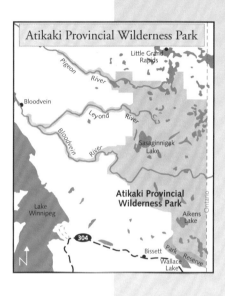

The Bloodvein River has many moods, offering the challenge of white water as well as the tranquillity of still stretches such as this one, left.

Central Manitoba

Central Manitoba
The Legacy of Lake Agassiz

The Central Manitoba Lowlands are largely the legacy of glacial Lake Agassiz, once the world's largest lake. Stretching from the Trans-Canada Highway nearly 700 kilometres northwest, they cut a broad swath between the Precambrian Shield and the Manitoba Escarpment. The region's lacustrine heritage is evident in its three great lakes – Lake Winnipeg, Lake Manitoba and Lake Winnipegosis – that continue to shape its character.

In certain places and at certain times, this is a summer playground, popular and populous, particularly along its waterfront edges. At other times and in many more places, Central Manitoba is an unfrequented sweep of field, wood or marsh, secluded, untamed and burgeoning with wildlife.

In Manitoba's Interlake – the region between the great lakes – the rural roads that crisscross the southern exposure give way farther north to meandering tracks. Travel PTH 6, one of the Interlake's main arteries, from Winnipeg north to the Precambrian Shield and you'll see the vegetation change from open prairie and marsh through aspen thickets and pasture land to mixed woods that give way to thick stands of conifers. It's a transformation of landscape unrivalled by Manitoba's other regions.

Few sites in this central slice of the province are far from water in one form or another. From south to north, the landscape is studded with marshes, fens, bogs, tiny lakes, streams and long shorelines. Elsewhere, there are other reminders of the enormous glacial lake, which, at various times during the retreat of the glaciers about 11,000 years

Impressive even at rest, this American white pelican, left, signals its readiness to breed with the appearance of a flat, rounded plate on its bill.

The Shoal Lakes – East, West and North – northwest of Winnipeg offer prime breeding habitat for shorebirds and colonial nesting birds such as pelicans and cormorants.

Central Manitoba

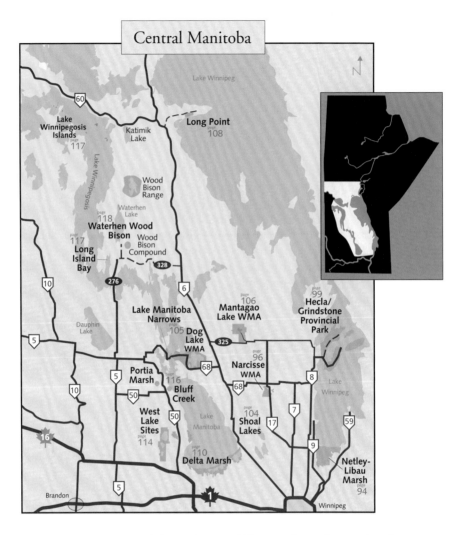

Photographers and birders at Delta Marsh, right, await the rising sun.

ago, covered the entire area. Eskers, end moraines and ancient beaches provide subtle relief in a mainly horizontal landscape.

Thanks to this abundance of water, Central Manitoba is one of the province's best places to view waterfowl. And its northern wilderness provides prime habitat for a range of ungulates, including elk, moose, white-tailed deer and a small but growing population of wild wood bison. This is also home to the world's largest garter snake hibernaculum.

The beaches may draw crowds for sun and sand, but much of Central Manitoba is an unbeaten path, with hidden treasures in unlikely places. For wildlife viewers, the rewards are ample.

Trip Planning

If you have a day in the region . . . ✪ Take your binoculars to Delta Marsh in the spring to see warblers and waterfowl. ✪ In late April or early May, don't miss the snakes of Narcisse WMA.

If you have two or three days . . . ✪ Drive a 500-kilometre loop counter-clockwise around Lake Manitoba's southern basin beginning at Delta Marsh and including Shoal Lakes, Lake Manitoba Narrows, Portia Marsh and Bluff Creek and the West Lake Sites. ✪ Drive a 300-kilometre loop from Winnipeg north up PTH 7 and 8 and return south along PTH 9. Along the way visit Oak Hammock Marsh WMA (featured in Winnipeg section) and Hecla/Grindstone Provincial Park.

If you have a week or more . . . ✪ Spend your time at Hecla/Grindstone Provincial Park or split your time between there and St. Ambroise Provincial Park at Delta Marsh. ✪ Travel the length of the Interlake on PTH 6 to Long Point, then head to Waterhen to see the wood bison. This journey could also be combined with a longer trip to northern Manitoba, which might include Pisew Falls or Grass River Provincial Parks (featured in the Northern Manitoba section).

Central Manitoba

Netley-Libau Marsh

Brian Wolitski

Though its parents are plumed in black and gray, this American coot chick looks like an iridescent ball of down.

Where the Red River merges with Lake Winnipeg lies one of the province's largest and most productive marshes. From the viewing tower at the End of Main, the farthest extent of Selkirk's (and Winnipeg's) Main Street, the delta stretches forth to the horizon like a northern bayou, a blend of river channels, winding creeks and shallow lakes.

Though this extensive marsh hugs the southern basin of Lake Winnipeg for almost 24 kilometres and lies just 45 minutes north of Winnipeg, Netley seems a well-kept secret compared to Manitoba's more renowned wetlands, Oak Hammock and Delta Marshes.

Challenging access may account for this relative anonymity, but herein lies its appeal. At its widest nearly 18 kilometres from south to north, Netley Marsh is big, open country, short on crowds, but rich in wildlife. For wildlife viewers, the rewards are considerable – once you gain access.

There are a couple of easy access points for casual visitors. The End of Main, where east-flowing Netley Creek joins the Red River at the north end of PR 320, is perhaps the most convenient, offering parking, restaurant facilities (in season), a campground and a boat launch as well as a viewing tower with interpretive signs. On the west side of the marsh, parking, dining, accommodations and a boat launch can be found six kilometres east of Petersfield, in cottage country north of the convergence of Netley and Wavey Creeks.

Deeper access requires driving on gravel or mud roads, followed by explorations on foot or canoe. Trails exist, but they're unmarked and can disappear in summer. In autumn they're more apparent, due in part to their use by hunters in search of Netley's abundant waterfowl.

Dikes constructed to control flood waters are often the best places from which to view wildlife. The most southerly of these is two kilometres north on Whiskey Ditch (a mud road despite its name), which crosses PR 515, 6.5 kilometres east of Clandeboye.

There are also several dikes on the west side of Netley Lake and Cochrane Lake, accessible by section roads east of PTH 9, north of Petersfield. On the east side of the marsh, drive 6.5 kilometres north of Libau on PTH 59, then another 2.5 kilometres west on the section road to the dikes between Lower Devil Lake and Swedish Lake.

Canoeing may be the best way to view Netley Marsh's wildlife bounty, but it can be dangerous. Netley's water flow is unpredictable, subject to wind tides and strong currents. The landscape can change rapidly. A south wind can flush water from the marsh like a plug pulled from a bathtub, leaving a canoeist suddenly high and dry. A north wind will flood the marsh and quickly open a confusion of unexpected water channels. It's easy to get lost if you aren't careful.

 This changeability gives Netley Marsh some of its special properties. It is renowned for its abundance of waterfowl, such as pied-billed grebes and American coots, that nest on floating vegetation. Diving ducks such as greater and lesser scaup use the marsh as a staging area in spring and fall. Shallow water feeders, including mallards, northern shovelers and blue-winged teal, nest near smaller ponds or sloughs in the upland areas.

In summer, Netley is a good place to view yellow-headed, red-winged and Brewer's blackbirds, warblers such as common yellowthroats and northern waterthrushes, marsh and sedge wrens, as well as song and swamp sparrows. Northern harriers are commonly seen patrolling above the reeds in search of voles.

The marsh also supports mink, beavers, white-tailed deer and large numbers of muskrats.

Jerry Kautz

Netley Creek flows through a maze of woods and wetlands toward its meeting with the Red River.

Access Netley Marsh is about 60 kilometres north of Winnipeg, framed by PTH 9 and PTH 59, but access is primarily by municipal roads and trails.

A good map is advisable. Topographic maps and Red River navigational charts are available from the Department of Natural Resources Land Information Centre (Map Sales). Novices might consider hiring a knowledgeable outfitter from the area as a guide or carrying a Global Positioning System instrument and a cell phone.

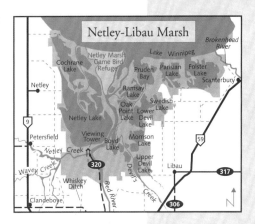

Central Manitoba

Narcisse WMA

In Manitoba, snakes and Narcisse are virtually synonymous in the public imagination. Every spring, tens of thousands of red-sided garter snakes and some western plains garter snakes slither to the surface from their winter dens in the limestone crevices in the Narcisse Wildlife Management Area and engage in a frenzy of mating.

This seasonal massing of hormonally-charged reptiles has earned Narcisse an international reputation for having the world's largest concentration of red-sided garter snakes. And this is so because conditions here are ideal for this species, which endures the coldest climate of any reptile.

Like much of the southern Interlake, the area around Narcisse is marked by underground caves and sinkholes carved by water from limestone that was once, hundreds of millions of years ago, the bed of an ancient sea. In most of southern Manitoba these limestone cavities, or karst formations, are buried beneath glacial sediment, but in parts of the Interlake region they are very near the surface. At Narcisse WMA, the roofs of these cavities have collapsed in several places, allowing access to underground chambers in which snakes hibernate.

The WMA has several active snake dens connected by a three-kilometre self-guiding interpretive trail. Though red-sided garter snakes are harmless, they are best viewed from observation platforms constructed adjacent to the dens.

Snakes are undoubtedly the WMA's great attraction, but there is other wildlife here. The area has 40 kilometres of other trails suitable for hiking or, in winter, cross-country skiing. Typical of much of the Interlake, the landscape consists of open grasslands interspersed with groves of aspen, ridges of bur oak and scattered white spruce.

The red-sided garter snake, so named for a series of red blotches along the side (the snake's dominant colors are black and yellow), engage in their annual mating rituals for a one- to three-week period sometime in late

Escaping the frenzied throng, above right, this large female red-sided garter snake is ardently pursued by a single-minded suitor.

Peter St. John

April or early May, depending on weather conditions. The males so vigorously pursue the larger thicker females that great balls of intertwined snakes, some containing hundreds of the creatures, form and reform, the sound of their skin rubbing together setting a hum like a tuning fork. Then they disperse to their summer feeding areas in the surrounding wetlands, where there is an abundance of frogs and toads to dine upon. Some have been recorded travelling as far as 17 kilometres. In the fall they return to their crevice homes in the limestone, perfectly located below the frost line but above the water table.

Snake-viewing season is brief, but Narcisse has other attractions. The WMA's mixed wood forest is a critical wintering habitat for more than a thousand white-tailed deer, as well as smaller numbers of elk and moose. Sharp-tailed and ruffed grouse are plentiful – the latter can be heard drumming on evenings in May and June. You may also be rewarded with sightings of red-tailed hawks, great horned owls and songbirds such as clay-colored, savannah and song sparrows, Swainson's thrushes, Baltimore orioles, yellow warblers and American goldfinches. Upland sandpipers also nest in the area.

> Snakes gather in such remarkable numbers at Narcisse WMA because of a combination of ideal winter and summer habitats. Deep caves suitable for hibernation are located near wetlands and shallow marshes, where the snakes live during the spring and summer months.

Access

From Winnipeg's Perimeter Highway (PTH 101), follow PTH 7 north 45 kilometres to Teulon, then travel an additional 45 kilometres northwest on PTH 17 to Narcisse. From Narcisse continue north for six kilometres. Access to the snake dens is on the east side of PTH 17 between Narcisse and Chatfield. Other trails are accessible from PR 419 west of Chatfield. The snake den section of the WMA is serviced with washrooms, water and picnic areas, and wildlife interpreters staff the sites during the viewing season. The remainder of the WMA is unserviced.

Central Manitoba

Hecla/Grindstone Provincial Park

An expanse of water stretches to meet the dome of the sky, rugged cliffs loom above sandy beaches, gulls and cormorants wait patiently as fishermen pull their nets from small boats. The scene might be any fishing hamlet in Atlantic Canada, but it's not. This is Hecla/Grindstone Provincial Park, in the heart of the Canadian prairies.

More water than land, the park comprises a large peninsula and a series of islands along the west side of Lake Winnipeg, the world's 13th largest freshwater lake. With wild spaces adjacent to recreational and resort developments, Hecla Island is very popular with visitors and harbors a variety of wildlife.

Aboriginal Manitobans have been drawn to Black Island for thousands of years; today the region's Anishinaabe gather there for annual berry-picking and social events.

Icelanders have called Hecla Island home since the 1876 eruption of Mount Askja forced many to leave Iceland. They emigrated to Canada and relocated on the island, the most northern part of a territory called New Iceland, which soon became a part of Manitoba. Here they fished and farmed for almost 100 years. Many had left by 1969 when Hecla was designated a provincial park, but fishermen continue to earn a livelihood from the waters around the island.

The park's forest habitat includes white and black spruce, balsam fir and tamarack, and mixed stands of spruce with aspen, balsam poplar, ash and white birch. Manitoba's most northerly stand of red pine is found on Black Island. Red-osier dogwood – dubbed moose candy – is prolific at forest edges on Hecla Island, as are large stands of willows and wild raspberry bushes. Hecla also has bogs and fens, and an extensive marsh adjacent to the causeway (see Grassy Narrows Marsh Trail).

Found throughout much of Manitoba, the common tern is an excellent fisher that captures its prey by diving from the air.

Great blocks of limestone, left, are typical of the shoreline at the north end of Hecla Island.

One of Hecla's success stories is the recovery of American white pelicans. Listed as "threatened" by the Committee on the Status of Endangered Wildlife in Canada in the late 1970s, the species has made a remarkable recovery. By 1987 nearly 50,000 birds were nesting in 70 colonies across Western Canada and the threatened designation was removed.

Hecla/Grindstone Provincial Park is really about birds – pelicans and cormorants, blackbirds and sparrows, warblers, wrens, swallows, woodpeckers, gulls, grebes, mergansers, cranes, herons, eagles, owls, geese, grouse and more.

During May and early June, western grebes, immersed in their intricate courting dances and oblivious to spectators, can often be seen from the causeway leading onto Hecla Island. For a view of this mating spectacle, pull off on the side of the road and focus your binoculars along Grassy Narrows, the channel that separates Hecla from the mainland.

In mid-October, bald eagles gather along the eastern coast of the island. For a fine view, follow the Lighthouse Trail to the end of the peninsula. The park held its first annual "Eaglefest" in the fall of 1998, featuring artists, ornithologists and a symposium on eagles.

Hecla Island and the Grindstone Peninsula are also known for their moose population. There are about 80 moose on the island, but daytime visitors may see scat or tracks rather than these large, rather reticent ungulates.

The park also supports a small population of wolves, and park interpreters feature two wolf-howling nights in August. Veterans of "howl connections" say they are just as thrilling as sightings.

The surrounding waters of Lake Winnipeg support 24 species of fish, including yellow perch, sauger, walleye, goldeye, sturgeon and whitefish.

Hecla/Grindstone Provincial Park is 165 kilometres north of Winnipeg on PTH 8. Hecla Island's size, (25 kilometres long, seven kilometres wide) makes it easy to get from one side to another and all trails are clearly marked and well-maintained.

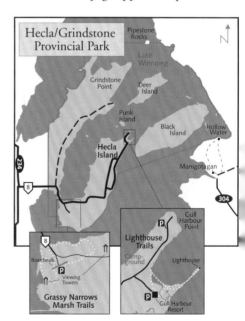

Grassy Narrows Marsh

This trail system is east of the Grassy Narrows channel, just over the causeway that links Hecla Island to the mainland. In recent years, Grassy Narrows Marsh has been rehabilitated by Ducks Unlimited, with water levels managed to optimize habitat conditions.

Dike paths and boardwalks meander through the marsh, with towers and blinds to enhance viewing possibilities. Trails, ranging from a half-kilometre to 10.6 kilometres, are clearly marked with wildlife symbols and colors. The two long trails are best enjoyed with the use of a rugged bicycle.

Grassy Narrows Marsh supports hard-stemmed bulrushes that grow to their maximum height of nearly two metres, and cattails – featuring huge brown seed heads – that extend a full 2.5 metres above the water. Giant reed grass or phragmites sometimes sways nearly four metres high.

Beginning as early as April and extending through the summer, this magnificent marsh teems with life. Sixty-two species of birds have been identified. Giant Canada geese nest here, and other subspecies of Canada geese are seen in migration. There are puddle ducks galore, including mallards, pintails, and shovelers. Common goldeneye, common mergansers, bufflehead and wood ducks nest in tree cavities around the marsh. Several members of the grebe family – western, horned, red-necked and pied-billed – make their homes here. Great blue herons may be seen wading in shallow waters and black-crowned night-herons are occasionally seen.

Other species include sandhill cranes, common and Forster's terns, soras and American coots. Caspian terns stop here during spring migration. Herring, ring-billed and Franklin's gulls make forays overhead in search of food. The marsh also attracts ospreys and three species of blackbirds. American bitterns are often flushed from the growth along the dike trails.

Apart from birds, the marsh supports beaver, mink and muskrats. Moose, or the distinctive cloven shape of their tracks

Grassy Narrows Marsh attracts more than 60 species of birds and boasts both hiking and cycling trails.

Photographer's Notes

June 15th: "Spent the evening along the lakeshore, waiting for the sunset and watching common terns. They dive from such heights, it's a wonder they don't break their necks. And when they did catch a meal, a pair of pesky pelicans were there to try to steal it."

As autumn approaches, bull moose remove the velvety covering of skin from their fully-developed antlers by rubbing them against trees or thrashing against branches. This majestic bull wears a leafy ornament that attests to his exertions.

along the shoreline, can sometimes be seen. Western painted and snapping turtles, blue-spotted salamanders, Canadian toads and red-sided garter snakes all live here. A dawn or dusk visit will likely be accompanied by a marsh frog chorus, involving leopard, wood and boreal chorus frogs.

Just over a kilometre inside the park entrance, a short trail on the north side of PTH 8 leads through a mixed forest to a wildlife viewing tower. This is a perfect location for birders and photographers. During spring and fall migrations, it's possible to see a great variety of species and large flocks of birds. At dawn or dusk a careful perusal might reveal a moose or a white-tailed deer. There are interpretive signs on the tower deck to help you in your search.

 The Grassy Narrows Marsh trailhead is located east of the causeway on PTH 8. The wildlife viewing tower parking area is one kilometre farther east on PTH 8.

Lighthouse Trail

Part of the Gull Harbour trail system, the Lighthouse Trail is wide and well-groomed. At three kilometres return, it is an easy morning or early evening stroll along the narrow peninsula that defines the northeastern tip of Hecla Island.

The trail meanders gently through the park's mixed forest of dense spruce, mature poplar and thick dogwood. Beyond and through the trees, photographers will enjoy many beautiful views of Lake Winnipeg. To the east, there's a fine view of Black Island. This part of the lake is also popular with boaters and sailors because of the protection afforded by adjacent islands.

Watch for common mergansers on the east side of the peninsula. They nest in tree cavities along the coastline and in summer the females, with their tufts of spiked auburn feathers, delight visitors with their parenting skills, as they herd up to 25 youngsters around the lake. Each hen merganser only hatches about 12 chicks, but some are seen with "gang broods" made up of the stray offspring of other commmon mergansers. Large concentrations of mergansers also gather here in late fall during migration.

Giant Canada geese also flock to this area. One pair, completely unruffled by human activity, usually nests on an island in a small pond adjacent to the resort.

Along the Lighthouse Trail you will also see pelicans, cormorants and with luck, bald eagles and ospreys. If you visit in mid-October, look for bald eagles in migration. These birds, with a wingspan of up to two metres, fly south past this point singly or in small groups of three to five. Also look for rough-legged hawks.

 Directions to the trailhead can be obtained at Gull Harbour Resort, located at the northern end of the island on PTH 8.

The channel east of Gull Harbour Lighthouse, above, between Hecla's northern tip and Black Island, can be treacherous when the wind blows from north or south, causing wind-driven tides.

The dark green head of the male common merganser often looks black. Like the western grebe, it can sometimes be seen running along the water before take off.

Shoal Lakes

Seen in large flocks during migration, the northern pintail drake can be identified at rest by its distinctive white neck markings and in flight by its long tail plumes.

The Shoal Lakes – East, West and North – provide important habitat for shorebirds of all varieties and important breeding grounds for colonial nesting birds. The area was favored by pioneer naturalist Ernest Thompson Seton. In *Birds of Manitoba*, he proclaimed "Shoal Lake" as a world-class nesting and breeding ground for pelicans and cormorants. What was once a single large lake during Seton's time is now three shallow, rather salty basins of water with little shoreline vegetation. These lakes continue to attract pelicans, cormorants, a great variety of shorebirds and waterfowl.

American avocets, willets, spotted and upland sandpipers, marbled godwits, common snipe and Wilson's phalarope breed along all three lakes. They are joined during spring and fall migration by semipalmated, white-rumped, Baird's and stilt sandpipers, and long-billed dowitchers. Greater and lesser yellowlegs, ruddy turnstones, dunlin and sanderlings can also be seen.

West Shoal Lake, an important nesting area for the endangered piping plover, has been designated a Western Hemisphere Shorebird Reserve Network site. These tiny birds are usually associated with coastal beaches. Here, they probe the sand and gravel bars for aquatic larvae, worms and small crustaceans and then flit to nearby grasses to search out insects and spiders.

North Shoal Lake attracts an uncommon number of red-necked grebes. In the surrounding grasslands, western meadowlarks, savannah, vesper, Le Conte's and clay-colored sparrows, horned larks and bobolinks nest in great numbers.

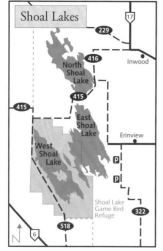

 Access From Winnipeg, travel 40 kilometres northwest on PTH 6 to Woodlands, turn north onto PR 518 along West Shoal Lake to PR 415, which runs between North and East Shoal Lakes.

Dog Lake WMA
& Lake Manitoba Narrows

Dog Lake Wildlife Management Area, situated just east of the Lake Manitoba Narrows in the central Interlake, is mainly marsh. Surrounded by private farms and ranches, its 13,000 hectares provide prime breeding habitat for Canada geese. The lake also serves as a major staging ground for all types of waterfowl during migration.

To the west, The Narrows link the north and south basins of Lake Manitoba. Constricting the flow of water, this is a place of wind-driven tides, where a north or south gale can send water thundering through the channel. An ancient legend says this almighty noise is the voice of Manitou, from which the name of the province is derived – Manitoba means "the place where the spirit lives".

Guy Fontaine

Dog Lake WMA is the only known nesting location of great egrets in Manitoba.

In addition to large numbers of Canada geese, white pelicans, ring-billed gulls, common terns and great blue herons breed in Dog Lake WMA. Islands in the lake are game bird refuges. Watch for great egrets, which nest in small numbers among the herons. In April and September, huge flocks of geese also stage here. This bounty – eggs and young, as well as some mature birds – attracts coyotes and foxes in large numbers. White-tailed deer can also be seen.

Giant Canada geese also nest at The Narrows, and rocky beaches and islands both north and south of the channel provide nesting habitat for ring-billed gulls, black terns, pelicans and double-crested cormorants.

Access: The Narrows is accessed by PTH 68 from PTH 5 west of the lakes or PTH 6 through the Interlake. PTH 68 also provides the closest access to Dog Lake WMA. From Winnipeg, take PTH 6 north through the western Interlake to PTH 68. Turn west and follow the highway to both Dog Lake and The Narrows.

Dog Lake Wildlife Management Area & Lake Manitoba Narrows

Photographer's Notes

June 13th: "Ditches are full of water along the road between the lakes. Saw pelicans, cormorants, lesser scaup, ring-necked ducks, meadowlarks, killdeer, a northern harrier, a red-tailed hawk, an eastern kingbird and a porcupine. You can get quite close to critters if you take your time."

Mantagao Lake WMA

The Mantagao Lake Wildlife Management Area encompasses more than 50,000 hectares of valuable wildlife habitat including Mantagao Lake and a significant portion of the Mantagao River that drains into Lake Winnipeg. The upland ridges include glacial eskers, ancient beaches and end moraines. Between them are cradled lush meadows and wetlands.

The aspen forest, stands of spruce and open meadows of the WMA provide prime habitat for elk. The WMA was established in 1967 to protect and augment the area's small remnant elk herd. Over the next six years, more elk were relocated – usually a dozen or more at a time – from Riding Mountain National Park.

Much of the area burned in a fire in the late 1980s. However, regrowth sprang up in the months following the blaze, and the process of forest succession is now well underway. A young aspen forest dominates, but a vigorous growth of jack pine has also reestablished itself. Several prairie and eastern forest plant species reach their northern limits in the area, including smooth sumac.

Ring-necked ducks, like this handsome male, are common in Manitoba's woodland ponds.

Moose, elk and white-tailed deer are found throughout the area, with some of the Interlake's best moose habitat along the Mantagao River. This is also one of the better places in Manitoba to see black bears. The non-poisonous northern redbelly snake is abundant here and may be seen slithering away from a trail. Beavers, mink and muskrats share the marshes with Canada

Elk are called intermediate feeders because in addition to grazing, they also eat the tender tips of young birches and willows in winter. Elk tracks are larger than deer tracks, smaller than moose tracks, and rounder than both.

geese, sandhill cranes and a great variety of ducks, including blue-winged teal, mallards, canvasbacks, redheads and lesser scaup.

Access From Winnipeg, travel north on PTH 7 to Teulon, then turn west on PTH 17. Travel north to the end of PTH 17 where it meets PR 325. Turn west on PR 325 for about 20 kilometres, then turn north on the road to Mantagao Lake.

Long Point
& Katimik Lake

From the air, Long Point looks like a giant finger jutting into the north basin of Lake Winnipeg. Nearly 40 kilometres long and up to 15 kilometers wide, it is the most easterly extension of a C-shaped glacial feature known as The Pas Moraine. Though the composite gravel, boulders and sand of a typical moraine formation can be clearly seen, the peninsula also supports hardy vegetation.

This is mid-latitude boreal forest, a mix of aspen, birch and bur oak, with black spruce in the lowlands and white spruce and jackpine along the high, dry ridges. Long Point is unusual in its abundance of eastern white cedar. Commonly found far to the southeast, they are near the northern edge of their range here and they are stunted, compared to their southern counterparts.

As is true elsewhere in the Interlake, there are many fens and bogs, identified easily by the presence of tamarack and black spruce. These dark, damp places provide perfect habitat for calypso orchids, northern green orchids and ram's head lady's slippers.

Long Point has extensive beaches on its north and south shores. On the south side, the land slopes steeply to a shoreline littered with rock and boulders. The north shore is more inviting, with long sandy beaches that wind in and around smaller spits and lagoons. Here, you'll also find sand dunes, shaped by wind and wave action.

Piping plovers depend in part on camouflage to protect their eggs, which are nearly invisible on a pebbly beach.

The sheltered bays and pebbled beaches of Long Point, particularly around Gull Bay, are perfect nesting areas for the endangered piping plovers. Visitors may be prevented from accessing this area during the nesting season.

American white pelicans are common here, as are bald eagles and ospreys. They all feed on an abundance of fish. There are also warblers, woodpeckers, and spruce and ruffed grouse along the forested ridge.

A sandspit curves southwest from the southern shore of Long Point, creating a quiet backwater. Piping plovers use the spit as a nesting site.

Local residents boast that this area is home to wildlife "ranging from a mouse to a moose". Black bears are plentiful, often in colors other than black. This locale has bears in all color phases, from black and various shades of brown to hues of cinnamon, blond and steel blue.

Ranges of four large ungulates – moose, white-tailed deer, elk and woodland caribou – overlap here, though the last two are rarely seen.

Katimik and Kaweenakumik Lake are located midway between Lake Winnipeg and Lake Winnipegosis off PTH 60. Surrounded by mixed wood forests of aspen and poplar, spruce and pine, they provide prime habitat for moose.

Katimik Lake is large and shallow, edged by cattails and bulrushes, sedges and tall grasses. In late September and October, the lake is literally covered with ducks, such as redheads and canvasbacks, lesser scaup and ring-necked ducks. It's also an important staging area for Canada geese, gulls, terns and grebes. Islands in Kaweenakumik Lake provide nesting habitat for colonial waterbirds.

Habitat diversity and abundant wildlife are among the reasons that both Long Point and Katimik Lake are being considered as part of a new Manitoba Lowlands National Park.

Access To reach both Long Point and Katimik Lake, travel 360 kilometres north of Winnipeg on PTH 6. Access to Long Point is just south of the junction of PTH 6 and PTH 60; Katimik Lake is 15 kilometres west on PTH 60.

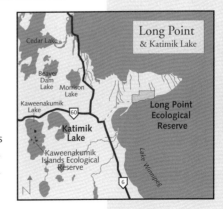

Delta Marsh

Delta Marsh is a vast wetland that rims the southern edge of Lake Manitoba, Canada's 14th largest lake. Created by the ancestral Assiniboine River, which once emptied into Lake Manitoba, this 18,000-hectare wetland is separated from the lake by a beach ridge that is more than 40 kilometres long.

Delta Marsh stretches from Lynchs Point east to St. Ambroise Provincial Park and beyond. East of the provincial park the marsh continues to the Lake Francis Wildlife Management Area, where tall grass prairie can be found in upland areas.

The lakeshore, with its sandbars and sand and pebble beaches, seems to stretch to the horizon. The beach ridge is often pounded by waves and bulldozed by ice, but huge cottonwoods, Manitoba maples, bur oaks and willows, as well as rare hackberry trees thrive here, attracting an amazing array of warblers during migration.

Extending up to six kilometres south, the marsh is a maze of waterways and bays, edged by bulrushes, cattails and reeds.

Instantly recognizable, the male yellow-headed blackbird is the only North American bird with a yellow head and a black body.

Like a bottleneck in one of the hemisphere's busiest migration corridors, Delta Marsh attracts hundreds of thousands of birds every spring and fall. Forster's terns nest in the marsh, while endangered piping plovers raise their young along the beaches. Common and Caspian terns are seen during migration.

Bird species from both lake and marsh habitats nest along the ridge, as well as woodland birds such as yellow warblers, least flycatchers, warbling vireos, Baltimore orioles and American goldfinches.

In the wetlands to the south, marsh birds, including sedge and marsh wrens, red-winged and yellow-headed blackbirds and the elusive American bittern, may be seen, as well as Wilson's phalaropes and great blue herons. Eared grebes, the males in their flamboyant rust and gold breeding colors, and western grebes both nest in the

Still arrayed in its breeding plumage, an eared grebe takes a turn at caring for the season's young.

Fast Facts

Roughly 10,000 songbirds were banded here during the summer of 1996 at the Delta Marsh Bird Observatory, one of the busiest bird banding stations in Canada.

The marsh is famous among waterfowl aficionados for its size, research traditions, hunting and even its art. The late ornithologist and artist, H. Albert Hochbaum, wrote *The Canvasback on a Prairie Marsh,* as well as other books, based on his observations of birdlife at Delta Marsh.

Two field stations are located at Delta Marsh. The one operated by the University of Manitoba offers weekend workshops to the public.

marsh and are seen on a regular basis throughout spring, summer and early fall.

At dawn in late April and early May, sharp-tailed grouse, in groups of as many as 20 males, may be seen performing their mating dances on leks along PR 411 at the southern edge of Lake Francis WMA. In June, these uplands provide prime habitat for Le Conte's and savannah sparrows and other grassland species.

It is during spring and fall migrations, however, that Delta Marsh truly earns its reputation as a birder's paradise. In April and May, and again in August and September, large numbers of warblers, sparrows, and other songbirds can be seen along the beach ridge and in October, tundra swans join throngs of migrating geese and ducks.

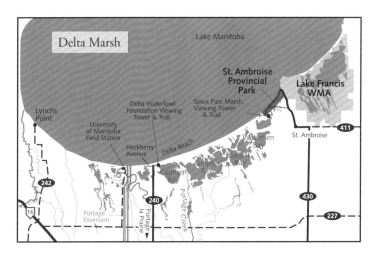

Central Manitoba

Portage Creek was so named because it provided a link between the Assiniboine River and Lake Manitoba, involving only a short portage. It runs into Simpson Bay east of Delta Beach. Fur traders carried or "portaged" their canoes a short distance over the prairie to the creek, then travelled through the marsh to Lake Manitoba.

East end: the half-kilometre Sioux Pass Marsh boardwalk is located at St. Ambroise Provincial Park day use area, at the end of PR 430 north of the Trans-Canada Highway (Number 1). Following the park sign, turn west on a gravel road about two kilometres north of PTH 411. Head north to the park entrance and turn left to the day use area. To access Lake Francis WMA, turn east at PTH 411. Less than six kilometres along, there is a small parking area on the north side adjacent to the tall grass prairie.

West end: hiking or cycling the levee along the Assiniboine River diversion gives elevated views over the marshland. From Portage la Prairie, drive north on PR 240 to Delta Beach. Turn west down Hackberry Avenue approximately three kilometres past the cottages to the north-south diversion. Woodland species may be seen in the cottage area and there are several locations where access to the lakeshore is possible as well. Alternatively, turn east just before Hackberry Avenue, named for the rare hackberry tree that grows along the southern margin of Lake Manitoba, and go over Delta channel, which runs between the marsh and Lake Manitoba. The Delta Waterfowl and Wetlands Research Station has a trail starting on the south side of the road just past the channel crossing and a boardwalk around a pond with viewing towers overlooking Cadham Bay. To access the bay, canoes may be put in at a channel along PR 240 less than 400 metres south of the town of Delta Beach. Watch for a parking area on the east side of the road. The University of Manitoba Field Station can be reached by driving along PR 227 to the Portage Diversion. Turn north along the road that follows the west side of the diversion and continue to the outlet at Lake Manitoba. Turn west along the access road to the station.

The courtship dance of a pair of western grebes, sometimes termed "racing", involves running over the water and finally diving.

West Lake Sites

Unlike many of its more flamboyantly-colored relatives, the pied-billed grebe is a rather solitary bird, which will generally escape by diving or slowly sinking below the surface of the water.

Three important natural areas lie within 25 kilometres of each other along a major migratory route west of Lake Manitoba. Langruth Wildlife Management Area protects some 2000 hectares of native grasslands, aspen groves and wetlands. Big Point is an extensive lakeside marsh along the western shore of Lake Manitoba. Big Grass Marsh, lying inland, is fed by runoff from the distant Riding Mountain escarpment via the Grass River.

Big Grass Marsh

This marsh complex is one of the most important waterfowl staging areas on the continent and was Ducks Unlimited's first restoration project in Canada (undertaken in 1938).

Enormous numbers of birds pass through this Game Bird Refuge, peaking around mid-June. This is also the best time to see some of the 5,000 Franklin's gulls nesting in colonies here. A drive on gravelled PR 265 affords views of ruddy ducks, American bitterns and pied-billed grebes.

More than 6,000 sandhill cranes stage here; they are easiest to see heading north in April and May, and returning south during September. More than 25,000 ducks come through in autumn, as well as hundreds of thousands of snow geese, often covering the marsh like a fluttering sheet. Red foxes scurry across the wet meadows, taking advantage of the bounty.

Though plans are underway for a viewing platform, fully experiencing the marsh requires a spirit of exploration. Take a dike trail on the east side of the dredged channel to the water control structure on the Grass River. Or, for best access, canoe through the tall reeds and grasses in spring and early summer. Put in at the bridge over the channel, about 13 kilometres west of Langruth along PR 265. Paddle north for a few minutes, portage over the dam, and continue on into Jackfish Lake.

Langruth WMA

Just 10 kilometres west of Lake Manitoba, the Langruth WMA consists of ridges and swales, mainly aspen woods and grasslands in the higher areas and wet meadows in low-lying spots. Used as a practice bombing range by the Royal Canadian Air Force during the Second World War, it now provides habitat for white-tailed deer, waterfowl and grouse.

On early mornings in April or May, visitors may be lucky enough to see a group of male sharp-tailed grouse performing a ritual courtship display on a dancing ground, called a lek. From about 45 minutes before sunrise and continuing for up to three hours, males coo, cackle and inflate their neck sacks while they stamp, strut and run with outstretched wings in an attempt to attract females.

Big Point and Hollywood Beach

Big Point juts out from Lake Manitoba's west shoreline, creating a prime spot to watch American white pelicans along the shore and Forster's terns diving for fish. Willets and marbled godwits are commonly seen feeding at the water's edge.

There are both marsh and sedge wrens, red-winged and yellow-headed blackbirds and a colony of between 200 and 300 western grebes.

"On May 23, 1884 ... heard a bittern, above, pumping in the slough after dark; the sound has been very aptly likened to the syllables *'pump-o-ga;'* the first two notes are like the stroke of a pump, the last is exactly like the swish and gurgle of water in a deep pipe."

– Ernest Thompson Seton, *Birds of Manitoba*

Access: From Winnipeg, travel 90 kilometres west on the Trans-Canada Highway. Turn north on the Yellowhead Highway and continue about 30 kilometres to PTH 50. Turn north about 26 kilometres to Langruth. Langruth WMA is five kilometres north and two kilometres west of Langruth. Turn west at the Hollywood Beach sign. The gravel road (PR 265) between Langruth and Plumas cuts through Big Grass Marsh, about 13 kilometres west of Langruth. Head 12 kilometres east from Langruth on PR 265 to get to the peninsula of Big Point. To access the north side of Big Point marsh, follow the road at the north end of Langruth WMA east toward Hollywood Beach and Lake Manitoba. Put a canoe in the ditch and paddle into the marsh.

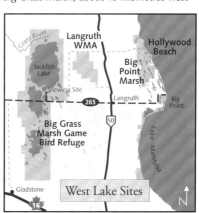

West Lake Sites

Portia Marsh & Bluff Creek

Ian Ward

South of Bluff Creek is an excellent place to view small yellow lady's slippers in early June, followed later in the month by showy lady's slippers, above.

The trails of Portia Marsh and Bluff Creek offer some of the best wildlife viewing on Lake Manitoba's west side, largely due to the efforts of the Alonsa Conservation District.

Portia Marsh

This wetland complex supports many bird species, and mammals such as mink, muskrats and beavers. At the trailhead, visitors will find an observation tower, marsh boardwalk and nature trail. Interpretive signs explain this wetland environment and identify many of its colorful inhabitants. The boardwalk connects with a trail that leads through an aspen forest on return to the parking lot.

Great blue herons, American white pelicans and American bitterns may be seen. During spring and fall migrations, watch for Canada and snow geese as well as canvasback ducks, redheads and blue-winged teal.

Bluff Creek Nature Trail

This three-kilometre trail winds along a creek, through aspen forest and loops around a marsh. White-tailed deer may emerge from the trees or a flock of pelicans may be seen scanning the creek for fish. Migrating warblers such as Wilson's, bay-breasted and Cape May warblers can be seen in May. This is also a particularly good site for pileated, hairy and downy woodpeckers as well as a small number of red-headed woodpeckers. Red-sided garter snakes are sometimes seen and, on rare occasions, black bears.

Access: From Winnipeg, travel 90 kilometres west on the Trans-Canada Highway to the Yellowhead Highway, past Portage la Prairie. Take the Yellowhead Highway to PTH 50. Follow PTH 50 past Amaranth to where it turns west to Alonsa. Instead of making the turn, continue north on PR 278 to the sign for Margaret Bruce Beach. Follow this road east to Lake Manitoba and the Bluff Creek trailhead, on the north side of the road near the community hall. For Portia Marsh, continue 12 kilometres along PR 278 to the signs for the marsh. Take the road leading west for four kilometres to the trailhead on the south side of the road. The best lady's slipper viewing is five kilometres north of Amaranth.

Long Island Bay
& Lake Winnipegosis Islands

From the base of Lake Winnipegosis, Long Island Bay points north into Manitoba's Interlake like the tip of an anchor. Accessed only by water, the 30-kilometre bay and its associated marshes attract large flocks of migrating ducks every fall.

Islands in the bay, like many others that dot the length of Lake Winnipegosis, are home to large numbers of colonial nesting birds. It is to protect these birds, as well as a population of melanistic (or black-colored) garter snakes, that a number of islands farther north have been included in the proposal for a Manitoba Lowlands National Park.

Drake redheads gather on Long Island Bay in summer to molt their flight feathers.

Long Island Bay is best known for its large number of drake redheads that gather on the bay in summer. Here the drakes molt their flight feathers in relative safety from predators, while their erstwhile mates and young have to contend with the flightless period on small wetlands far to the southwest in the Minnedosa pothole country. American white pelicans, double-crested cormorants and great blue herons are also common on the bay. Herring gulls and ring-billed gulls are abundant, with smaller numbers of California gulls, at the northeastern edge of their breeding range. Caspian terns can also be seen on islands in the bay and along the ragged shorelines.

A number of islands and rocky reefs farther north, where the great lake curves west beneath The Pas Moraine, support some of the greatest concentrations and diversities of inland colonial nesting birds anywhere in Canada. Others, near the northwestern tip of Lake Winnipegosis, serve as denning sites for a rare black subspecies of plains garter snake. In the spring, the snakes emerge from caves and fissures below the frost level and swim to the mainland to spend the summer in marshes and meadows along the shore.

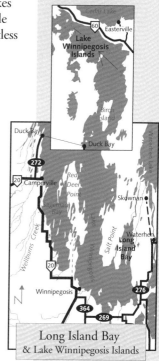

Both Long Island Bay and the islands of Lake Winnipegosis are accessible only by water.

Long Island Bay & Lake Winnipegosis Islands

Waterhen Wood Bison

Bordered on the west by Lake Winnipegosis and on the east by PTH 6, the northwestern Interlake is an 8,000-square-kilometre wilderness of forests and fens. The upland aspen-oak ridges merge northward into thick spruce forests, while the lowlands mix meadows and marshes.

The thin, stony soil was deemed unfit for farming when Manitoba was settled, but for wildlife, this is a bountiful place, and for wood bison, it seems ideal.

In 1992, for the first time in more than a century, wood bison calves were born in the wild in Manitoba. A small herd of 12 Manitoba-bred animals had been released at Chitek Lake in March of 1991. The birth of the cinnamon-colored calves the following spring was the first in a series of successes for the Waterhen Wood Bison Project. In the years since, the herd has grown to more than three dozen animals.

The herd is the culmination of a joint project by the Manitoba government and the Waterhen First Nation, part of an international effort to restore wood bison to the wild. Larger, darker and less numerous than their plains bison cousins, wood bison faced extinction by the end of the 19th century.

The Waterhen people were approached in 1984 to develop a captive breeding herd to produce animals that could live in the wild and not stray far from the release site.

The Chitek Lake herd is the only free-ranging population south of Canada's northern territories. Its presence makes this one of very few places in the country that supports white-tailed deer, moose, elk, woodland caribou and bison. The region is also home to wolves and black bears, raptors by the hundreds and waterfowl by the thousands, as well as a myriad of other species.

Wood bison cows will often not conceive if conditions are less than optimum. Food shortages, stress and disease can all cause declines in calf production. At Waterhen, however, the cows in both the wild and captive herds are healthy and apparently content, regularly producing calves and even occasionally twins.

The Waterhen River, left, links Lake Winnipegosis with Waterhen Lake and then empties Waterhen Lake into Lake Manitoba. Both Waterhen Lake and River are named for American coots, colloquially known as "waterhens".

Access There are no roads into the Chitek Lake area, but access is possible by snowmobile in winter. The captive herd can be viewed by prior arrangement with the Waterhen First Nation herd manager. The bison are located in a compound near Mallard, west of PTH 6, north of PR 328. Waterhen can also be reached by taking PTH 5 from the Trans-Canada north to Ste. Rose du Lac. Continue on PR 276 to Waterhen. Alternatively, take PTH 6 from Winnipeg northwest to St. Martin Junction and turn west on PR 328.

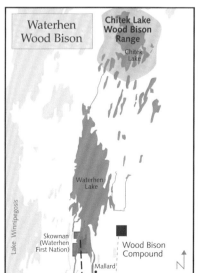

Southern Manitoba

Southern Manitoba
The Grasslands

Those most travelled of the province's highways, the Trans-Canada west of Winnipeg to the Saskatchewan border and PTH 75 south to the United States, very nearly frame southern Manitoba. But the view from these asphalt ribbons, unravelling as they do down the path of least resistance through the seemingly everlasting plain, gives only a bare suggestion of the complex and enchanting landscape beyond and out of sight.

Venture off these main arteries and you enter a realm of fields, woods, marshes, slender lakes, sinuous rivers and benign hills, interrupted here and there by touches of the dramatic – the steep-sided ravines along the Souris River, or the shifting dunes of Spirit Sands.

Perhaps no part of the province enjoys so profound a legacy of the ice age as does southern Manitoba. This was the land that first emerged from beneath the ice sheets of the Late Wisconsin glaciation, which lasted from 25,000 to 12,000 years ago. Though the underlying structure of the land was fundamentally the same after the glacial advance as it was before, the grinding glaciers and torrential rivers of meltwater modified the landscape in ways that are unmistakable today.

The hills south and east of Brandon are made of debris left behind by the ice when it stalled in its retreat. Bone Lake, Pelican Lake, Rock Lake and Swan Lake, strung like a necklace along the

White-tailed deer abound throughout Manitoba's southern grasslands and June is an excellent time to see young, like the tiny fawn opposite, almost anywhere in the region.

Southern Manitoba

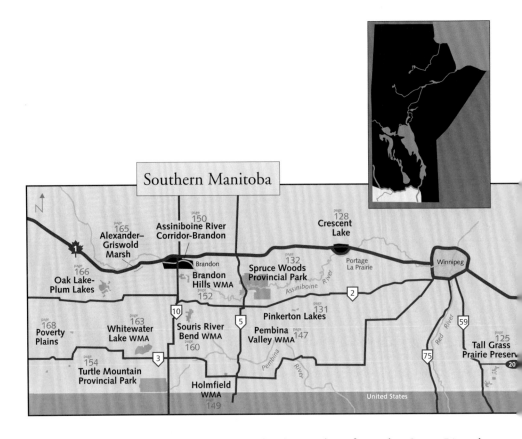

Pembina Valley, are placid reminders of a mighty Souris River that once carved its way southeast through the landscape. Spirit Sands is a remnant of an enormous delta where the ancient Assiniboine River flowed into glacial Lake Agassiz, a meltwater reservoir that once covered much of Manitoba. The flat fertile Red River Valley was once Agassiz's vast central basin.

In the wake of the glaciers, a great diversity of plant life evolved in this modified landscape. Through the Red River Valley, lush tall grass prairie grew, the northern tip of a sea of grass that once stretched south all the way to Texas. In the hilly land south of present-day Brandon, huge herds of bison grazed mixed grass prairie. Ribbons of cottonwood, Manitoba maple, green ash and American elm trimmed the meandering rivers, while marshy wetlands dotted the landscape.

Though substantially altered by human endeavor, some significant remnants of each habitat remain. And wherever habitats are diverse, so, too, is wildlife.

Southern Manitoba draws birders from near and far every spring, opposite.

Trip Planning

If you have a day in the region . . . ❁ For birds and flowers, visit the Tall Grass Prairie Preserve, a remarkable remnant of an ecosystem that has all but disappeared. ❁ In winter, try cross-country skiing at Spruce Woods Provincial Park. ❁ Discover the hidden birding delights of the sprawling Pembina Valley.

If you have two or three days . . . ❁ Spend your time in Spruce Woods Provincial Park. ❁ Visit the Brandon area – its Assiniboine River Corridor, or nearby Brandon Hills, Alexander-Griswold Marsh and Oak Lake-Plum Lakes. ❁ Hike, bike or ski at Turtle Mountain Provincial Park. ❁ In May or June, visit Poverty Plains in the province's southwestern corner to see grassland birds and stop at the bountiful wetlands at Oak Lake-Plum Lakes to see dozens of species of waterfowl and shorebirds. ❁ In late spring or early summer, paddle the Assiniboine River from Stockton ferry to PTH 34.

If you have a week or more . . . ❁ Drive a loop west from Winnipeg to Manitoba's southwestern corner that allows access to every site in this Southern Manitoba section, with the exception of the Tall Grass Prairie Preserve (featured in an Eastern Manitoba driving loop). From Winnipeg, drive west on PTH 3, north on PTH 21 and back to Winnipeg on either PTH 2 (for Souris River Bend, Brandon Hills and Pinkerton Lakes) or the Trans-Canada Highway (for Alexander-Griswold Marsh, Assiniboine River Corridor and Crescent Lake). ❁ For a more leisurely week, explore Turtle Mountain or Spruce Woods Provincial Parks or split your time between the two.

Southern Manitoba

Tall Grass Prairie Preserve

Ian Ward

Within one human lifetime, the prairies have passed from wilderness to become the most altered habitat in this country and one of the most disturbed, ecologically simplified and overexploited regions in the world. The essence of what we risk losing when the grasslands are destroyed is not a species here or a species there, but a quality of life, the largeness and wildness that made this country remarkable.
— Dr. Adrian Forsyth

This 2,200-hectare preserve protects a tiny remnant of a vast sea of tall grass prairie that once extended from Texas to southern Manitoba. Comprising three blocks of land in southeastern Manitoba, the preserve gives visitors a window on a past that has all but disappeared.

By the late 1800s, the region's rich black soil – a legacy of glacial Lake Agassiz – drew settlers by the thousands. In decades, they transformed seemingly endless stretches of grasses and wildflowers into fields of wheat. Much of the land near the Manitoba Tall Grass Prairie Preserve was settled by people from Bukovinia in western Ukraine. Homesteads and St. Michael's Church, Canada's first permanent Ukrainian Greek Orthodox church, can still be found in the area.

The 1.6-kilometre Prairie Shore Self-guiding Interpretive Trail is an inviting ramble through big bluestem and open sedge meadows. Along the way, visitors can view more than 150 plant species, a glimpse of Manitoba's once-remarkable prairie diversity. Most of the preserve is accessible year round for hiking and walking. If you stop to rest on a rock – or "sleeping sheep" as the locals call them – think of this: the presence of these large stones, brought here by the vast Laurentian ice sheet, saved this diversity from the plough.

Stretching up to a half-metre tall and boasting 10 or more intensely fragrant blossoms, the western prairie fringed orchid, left, can hardly be termed a wallflower. Its presence in Canada was unknown until the mid-1980s, when it was discovered clinging to existence along the ditches of a road in southern Manitoba. The survival of this rare and beautiful orchid was one of the reasons the Manitoba Tall Grass Prairie Preserve was created in 1989.

The monarch butterfly, opposite, on a meadow blazing star, can have a wingspread of up to 10 centimetres.

Sandhill cranes breed every summer in wetlands in and around the preserve.

The parade of colors changes almost weekly, making each visit a new experience.

From May to October, colors and fragrances change weekly. First, early blue violets and wild strawberry blossoms appear, followed by yellow stargrass, golden Alexander, blue-eyed grass and the medicinal Seneca root. In late May, small white lady's slippers, an endangered species, can be found in the wet meadows of the southern part of the preserve. In July, another endangered species, the western prairie fringed orchid – a creamy-white perennial with fringed lower petals – blooms, the only place in Canada that it does. Black-eyed Susans, purple prairie clover and prairie lilies add their feast of color to the mix. In August, there are more delights. Culver's root blooms pinky white, one of the few locations where it's found in Manitoba. Wild sunflowers follow the sun with their chocolate-scented faces. Golden Indian grass, big bluestem – reaching a height of two metres – and the very rare great plains ladies'-tresses enhance the scene. Once the gentians bloom and the monarch butterflies begin their migration south, we know that fall is here.

This is an excellent place to see birds, butterflies, frogs and beetles. White-tailed deer and moose are sometimes seen. Hikers are not likely to see many small mammals, but ubiquitous black mounds of earth reveal the existence of many northern plains pocket gophers. Shafts of moving grasses may mean that harmless plains garter snakes or smooth green snakes are passing by. Though the preserve was established to protect native plant species, this is also an excellent place to find songbirds. More than 50 species of birds nest in the preserve, while nearly 75 others have been sighted here. Visitors have a very good chance of seeing and hearing sandhill cranes, with their croaking cries, on almost any visit in spring through fall. Listen also for the bobolink, which emits a bubbling, gurgling song. This is also a good area to see many of the grassland sparrows, such as Le Conte's and the rare Nelson's sharp-tailed sparrow. Other interesting birds on the preserve are the upland sandpiper and the common snipe, which during its courtship flight display, makes a winnowing sound from the air moving through its feathers. Listen too for the sedge wren and the soft tapping of the yellow rail. Sharp-tailed grouse may also be seen. By July, more than 20 kinds of butterflies can be found. The first of these, the mourning cloak, appears while there is still snow on the ground. Monarch butterflies arrive in May and flit from milkweed to milkweed. An extended arm – held long enough – may attract the salt-loving yellow swallowtail. The rare, indeed unique in Canada, Powesheik skipper may be seen here. Early morning, when butterflies are cold and not yet moving, is the best time for watching or taking photos. Also, plenty of damselflies and dragonflies can be found here, the latter whirling like patrolling helicopters as they search for mosquitoes. During light rainy weather in early spring, visitors will hear the garbled quack of boreal chorus frogs or the banjo-like *clung, clung, clung* of the rare green frog. The leopard frog's call is similar to the sound of rubbing a wet balloon, sometimes followed by a chuckle.

Brian Wolitski

The common snipe, an inland sandpiper, above, generally stays close to cover.

Access From Winnipeg, go south on PTH 59 for about 90 kilometres, then turn east on PR 209 at Tolstoi for 3.2 kilometres. To access the interpretive trail, enter the parking area on the south side of the highway. Even insects are protected here, so look, and enjoy but don't dig, pull or pick any flowers or plants, or trap any insects. For more information, contact the Department of Natural Resources or the Manitoba Naturalists Society.

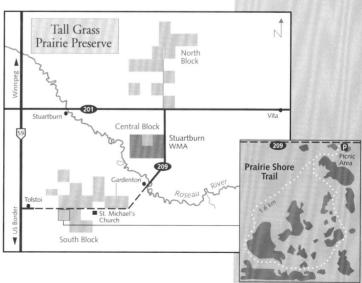

Crescent Lake

Dozens of pairs of mallards, Manitoba's most common and most recognizable duck, raise young at Crescent Lake each year.

Paleochannels east of Portage la Prairie mark the former courses of the Assiniboine River and can be seen clearly from the air, right.

As it has for thousands of years, the Assiniboine River snakes across Manitoba's southern prairies, uncoiling from west to east. But the meandering route it now takes to its junction with the Red River in downtown Winnipeg is not the path it chose in the past. The evidence of previous channels is clear in the multitude of oxbow lakes that parallel today's stream and in the ancient paleochannels that can still be seen from the air (see photo below), south of the present course.

The souvenirs of the river's meanderings occur when floodwater cuts a new channel through the neck of a winding loop, isolating the old river bend. Over time, these crescent lakes slowly vanish, turning first to marshes and later to wet meadows. During all these stages, oxbow lakes provide prime habitat for many species.

One of the best places to view an oxbow is in Portage la Prairie, where Crescent Lake encloses a 120-hectare peninsula called Island Park. Precisely when the oxbow was formed is not certain, but it was there in 1858 when Canadian surveyor Henry Youle Hind first laid eyes on it. "The old water course of the Assiniboine, near Prairie Portage," he wrote, "is now a long narrow lake, fringed with tall reeds, a favorite haunt of wild fowl and grackle, among which we observed the showy yellow-headed blackbird."

Just 40 years later, the lake was drying up. By the 1890s the burgeoning population of Portage had dubbed it "The Slough", and sections of it were being used as "hayland", a sure sign the water

was disappearing. Island Park had become enormously popular; by 1896 it was the site of a race track, grandstand and horse stables, as well as a golf course, a park and several farms.

These varied activities had so depleted wildlife that concerned citizens created a sanctuary and in 1912 flooded The Slough and renamed it Crescent Lake. The lake averages a metre deep, but is nearly twice that in some places. In its shallow southern extensions large cattail stands grow and water lilies can be found along the southeast corner of the bridge to Island Park.

The sanctuary today includes an eight-hectare floodplain forest of basswood, bur oak and green ash, with a few American elms. Beneath these large trees, American hazelnut, red-osier dogwood and highbush cranberry can be found, along with 20 rare alternate-leaved dogwood trees, a species at the northern edge of its range. Though on private land, the area is open to the public.

The rejuvenated oxbow is a haven for wildlife in an urban setting.

The combination of sanctuary and water encouraged the return of wildlife to the area soon after 1912. By the 1920s, Canada geese again were nesting along the shore and common goldeneye had taken up residence in big old American elms near the water's edge.

Today, many of the riverside elms are gone, victims of Dutch elm disease, but of 20 nesting boxes surveyed in 1997, 18 were inhabited by common goldeneye. Wood ducks find nesting cavities in some of the large cottonwoods, green ash and maple in the floodplain forest. Pintails and mallards nest along the lake and in neighboring yards; in autumn, large numbers of redheads and lesser scaup, along with hundreds of grebes and cormorants join them.

The Portage region is known for its magnificent trees, above.

Small numbers of common loons also stage here en route to their southern wintering grounds.

Crescent Lake's various habitats – riverbottom forest, parkland, lakeshore, wetlands and ponds – attract a variety of birds, both as residents and migratory visitors. Seasonal counts regularly produce more than 100 species, including nearly a dozen different species of warblers. Yellow-headed blackbirds still nest here, just as they did in 1858.

White-tailed deer are often seen, along with both red and gray squirrels. Frogs (particularly leopard frogs), toads and gray tiger salamanders thrive here, the last of these to the point of overabundance. Every few years, salamanders disperse en masse from the lake. Up to 30 centimetres in length, they undertake a fall journey to find new homes with a plodding determination that renders driving hazardous on Crescent Road and fills the window wells and backyard pools of adjacent homes with expatriot amphibians.

Access Crescent Lake is situated in the heart of Portage la Prairie, just a few blocks south of Saskatchewan Avenue (Highway 1A), the city's main street. It is the largest of three oxbows located within or immediately adjacent to the city limits. Crescent Road traces the lake's perimeter on the north side, circling past some of the community's fine old homes and leading to the island bridge. Particularly in autumn, large numbers of migrating waterfowl can be easily viewed from many places along the lakeside drive (although birders should take care to park on side streets, since Crescent Road is quite heavily used).

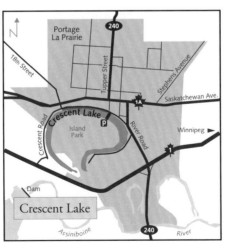

Pinkerton Lakes

Pinkerton Lakes lie on the northeastern edge of the Tiger Hills south of Treherne. There are two lakes, both long and narrow, connected by a channel. Slightly more than a kilometre from end to end, they seem cradled by the rolling farmland.

Dense emergent vegetation frequented by red-winged blackbirds fringes the shores. The songs of marsh wrens and sedge wrens can be heard from the reeds as you approach the lake. A high viewing tower overlooks the eastern shore; in May and October, it reveals the many species of birds that use the lake as a staging area during migration. Between May and September, Canada geese, blue-winged teal, ruddy ducks and mallards nest here and raise their young. Pelicans patrol the waters and great blue herons lift off in regal flight. Putting a canoe in here would be a great way to get closer to the birds, but not so close as to disturb them.

A trail begins about 50 metres south of the parking area. In winter skiers or snowshoers may see white-tailed deer, woodpeckers, white-breasted nuthatches or occasionally a great horned owl along this wood-lined trail.

Blue-winged teal, like this solitary drake, are common breeders on many of Manitoba's southern wetlands.

Access
From Winnipeg, travel on PTH 2 to the outskirts of Treherne, a distance of just over 100 kilometres. Turn south on PR 242 for about eight kilometres, up and over a hill. A sign and parking area on the right indicates the Pinkerton Lakes area. The tower stands some 20 metres from the parking area.

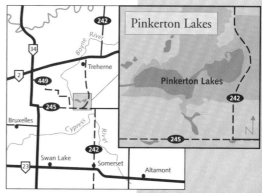

Spruce Woods Provincial Park

The famous pioneer-naturalist Ernest Thompson Seton loved the Spruce Woods area and spent many hours roaming the hills and thickets, notebook and sketchbook in hand. In August 1883, preparing to leave southwestern Manitoba, he wrote: "My time in this land is drawing to a close. I must start soon but the idea of leaving these fair flower starred bird enlivened prairies for horrible crowded smoky miserable New York is simply loathsome ..."

Spruce Woods Provincial Park encompasses almost 250 square kilometres south of the Trans-Canada Highway near Brandon. As Seton discovered more than 100 years ago, it is rich with unique flora and fauna.

The presence of some plants and animals is the result of current climatic conditions; others are relics of times long past. The combination makes Spruce Woods one of the most intriguing parks in Manitoba. Here you will find huge sand dunes rising 30 metres above the surrounding prairie, shifting and undulating in a timeless dance that has spanned 12,000 years. Here also are unusual reptiles, uncommon birds and rare flowers. Within the park are 75 kilometres of hiking trails; in winter they convert easily to ski trails attracting cross-country skiers from around the province.

The park's natural wonders are set in a coniferous forest of white spruce, which opens to stands of aspen and oak and wide tracts of native mixed grass prairie. Meandering through is the Assiniboine River, which continues to play an important role in the life of Spruce Woods.

The sand dunes and the river reflect the glacial legacy of Spruce Woods. About 13,000 years ago, the Assiniboine was a colossal river, two kilometres wide, rushing across the western plains to empty its torrential load of glacial meltwater into Lake Agassiz. An enormous delta of sand and sediment built up at the mouth of the river, until finally, in another age, it was left high and dry, fanning out over 6,000 square kilometres.

On warm days, turkey vultures soar on wings nearly two metres wide. The adults are easily distinguished from eagles by their reddish heads and two-toned wings.

Ladders provide access and protect the unstable dune faces, opposite, in Spirit Sands. Because the dunes are always in motion, park staff is kept busy changing the ladder placement.

Ernest Thompson Seton

Though born in England, raised in Ontario, and trained in France, Ernest Thompson Seton claimed it was Manitoba's Carberry Hills that had the greatest impact on his life's work as a naturalist, writer and artist. Seton arrived in Carberry in 1882 at the age of 21. There, captivated by the Spruce Woods area, he kept extensive notes on animal behavior and habitat, complemented by illustrations and paintings. These became the basis for several of his books, including *The Trail of the Sandhill Stag*.

Seton was named Manitoba's first provincial naturalist in 1892 and later founded the League of Woodcraft Indians, an international youth organization. In 1910, he joined Lord Baden-Powell in establishing the Boy Scouts of America. Seton eventually settled in New Mexico, where he died in 1946.

Southern Manitoba

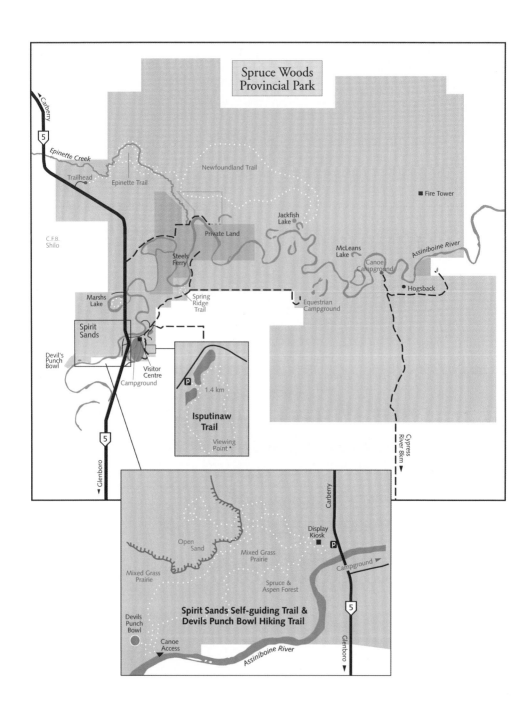

Of this once-vast delta area, just four square kilometres – called Spirit Sands – remain as open dunes that still move and shift restlessly with the prevailing northwesterly winds. This is demanding habitat even in summer; the heat of the open dunes doesn't provide a welcoming home for many animals or plants, save the hardy wolf spider and small stands of skeletonweed, sand bluestem and wolf willow, plants that cling to life on the sandy fringes, sucking moisture from a wealth of water underground.

The habitat on the valley floor 50 metres below the dunes, along the river, provides a striking contrast. Here is one of the longest tracts of protected riparian forest in southern Manitoba; great stands of ash, elm, basswood and cottonwood thrive in the damp soil, along with Manitoba maple and willows.

All over Spruce Woods underground springs feed wetlands and bogs, which in turn provide small, perfect habitats for plants that love dark and damp.

Spruce Woods has one of the largest examples of mixed grass prairie habitat in the province. Big and little bluestem, porcupine and spear grass, fescue, June grass – over 15 species of grasses can be found here. In spring, crocus and three-flowered avens create great sweeps of mauve and pink. Summer meadows are aglow with hairy golden asters, soft yellow prairie buttercups or deep yellow hoary puccoon. In winter, the prairie landscape becomes a dramatic duotone of white and tan against the dark green backdrop of white spruce.

Spruce Woods' natural diversity provides habitats for species with needs as varied as western painted turtles, above, and plains prickly-pear cactus, below.

The rolling hills and picturesque scenery of Spruce Woods draw cross-country skiers and snowshoers in winter.

The park's 12 oxbow lakes – formed from loops of the Assiniboine cut off from the main channel and nourished anew by runoff and underwater springs – create unique water communities, complete with 19 species of fish including the uncommon shiner dace and golden dace.

Spruce Woods gets such rave reviews about its unique landscape, the animals are sometimes overlooked. They are here in abundance, though, from large mammals to tiny reptiles. Majestic elk graze on the lush prairie grasses, white-tailed deer are commonly seen and moose have been sighted on occasion. Mink and raccoons can be found searching for food on the river sandbars, while the wetlands provide a perfect habitat for beavers and muskrats. There are foxes and coyotes (the latter can often be heard in a chorus at night), red and flying squirrels, ground squirrels, chipmunks, woodchucks and mice.

Two of the most intriguing residents in the park are the plains hognose snake and Manitoba's only lizard, the northern prairie skink. The hognose snake can sometimes be seen along the Spirit Sands trail, but it's not dangerous and if threatened will feign death.

Nobody quite knows how the northern prairie skink got here. Its closest relatives are in the United States, so it's assumed that, like the dunes and the forest, it's been left behind from another time. The little lizard lives throughout the park, but is hard to find; it's only about as long as a pencil and moves fast.

One of the most dramatic avian species in the park is the turkey vulture, circling the dunes in a lazy search for carrion. Broad-winged hawks and, occasionally, the northern goshawk can also be seen.

Visitors in spring and summer are likely to see bluebirds, particularly at nest boxes along PTH 5 in the park. Ruffed and sharp-tailed grouse can be seen year round, or at the very least, their tracks in winter.

In the coniferous forest are chickadees, gray jays, siskins and golden-crowned kinglets; where cones are bountiful, red and white-winged crossbills may be seen. Birds in the park's prairies include bobolinks, western meadowlarks and Sprague's pipits, the last of these more often heard above than seen. In spring, the river and oxbow lakes abound with wood ducks and mallards. Nesting boxes near the campground are also used by eastern screech-owls and northern saw-whet owls. Red-necked grebes and belted kingfishers are common summer residents of these wetlands as well.

Access Less than two hours from Winnipeg, Spruce Woods Provincial Park can be accessed from PTH 5 by travelling west on either the Trans-Canada Highway (Number 1) or PTH 2. From Brandon, go east about 40 kilometres to PTH 5, then turn south for 28 kilometres to the park entrance. The park campground is on the east side of the highway; the parking lot for Spirit Sands is on the west. The park abuts Canadian Forces Base Shilo on the west and when visitors imagine they hear thunder on a clear summer day, chances are it's the thunder of guns engaged in artillery practice. Those interested in the legacy of Ernest Thompson Seton might plan a side trip to the Seton Centre in nearby Carberry. **Canoe Access:** see Assinboine Canoe Route, this section.

Friends of the Bluebirds

Thanks to the efforts of a dedicated group of about 100 Manitobans, mountain bluebirds can now be found all over southwestern Manitoba and in many places along the west side of the province, north to Swan River. The Friends of the Bluebirds have installed and annually maintain nearly 3,500 nest boxes; most are placed on fence posts along roadsides.

Over the past decades this program has attracted hundreds of mountain bluebirds to Manitoba. This is the only province where the ranges of eastern and mountain bluebirds overlap. The best time to view bluebirds is between May and July, when they are busy building their nests and rearing their young.

Glen Suggett

Isputinaw Trail

This is the closest trail to the Spruce Woods Interpretive Centre and easy access alone makes it popular. But the fact that it's also a short (1.4-kilometre) loop through unusual habitats adds greatly to its appeal.

The trail climbs the southern slope of the valley carved by the ancient Assiniboine River, wanders along the top and descends once more to the river bottom. A self-guiding brochure explains the habitats encountered along the way.

Beginning with a boardwalk over a swampy area (be on the lookout for alder flycatchers; this is one of their favorite places in the park), the trail leads into a forest of ash, American elm, Manitoba maple, aspen and red-osier dogwood. Look for marsh marigolds in wet areas, as well as wood ducks in spring.

Soon the trail ascends the southwest-facing slope. The soil is drier here and in the few spots where the sun manages to penetrate the dense growth, golden Alexander and poison ivy take advantage of the bright warmth. Just above the valley floor hazelnuts abound. These nutritious nuts were gathered in late August by aboriginal Manitobans; later, settlers saved them for Christmas treats. Today they provide winter food for squirrels.

Common sense tells us the higher the slope, the drier the soil, but again Spruce Woods proves the exception. Water ripples to the surface in many places in the park, from valley slopes to sandy dunes. Here, springs have created a hillside bog, replete with sphagnum moss; in its centre a cluster of chokecherry bushes surrounds a dramatic, solitary white spruce.

Marsh marigolds brighten wet patches along the trail in late May and June. Look for them along the boardwalks that carry hikers over boggy areas.

From the heights of the Isputinaw Trail, hikers can see out over the broad Assiniboine Valley.

About halfway up the valley slope, the conditions favor a thick grove of bur oak, tightly entwined with wild grapes. After collecting hazelnuts, early Manitobans would continue on up to this plateau and gather large baskets of the wild fruit to dry for winter food or preserves. If this is where you catch your breath (and there are benches along the way), a quiet rest might be rewarded with the sight of a cedar waxwing or a ruffed grouse picking at wild grapes or chokecherries.

Moving higher, the soil is drier and bur oaks here are stunted compared to their taller cousins below. Tenacious ground junipers and bearberry spread upward, toward majestic white spruce that stand tall atop the slope. The view from the top is worth the climb – the Assiniboine River Valley winds to the west. Across the way is the densely forested northern valley, and in the distance, Spirit Sands.

The climb down to the trailhead takes a less dramatic route, passing through groves of silvery wolf willow with its lovely aroma.

Though perfectly colored to blend into its environment, the ruffed grouse gives away its presence in spring with loud drumming to attract a mate.

Spirit Sands- Devil's Punch Bowl Trails

For millennia, people have been drawn to Spirit Sands, attracted by the area's landscape with its unique plant and animal life.

Trail loops begin at the parking lot near an interpretive kiosk that reveals the area's history. An extensive trail system offers a variety of lengths and terrain. Hikers can begin with a trip to the dunes, detour to the Devil's Punch Bowl and return to the trailhead without retracing their tracks. The extended trip totals about 10 kilometres, but visitors can opt to spend a couple of hours exploring the dunes. It is advisable to carry water on these hikes.

The trail to the open dunes winds mainly through vegetated dunes, where the sand has been stabilized by plants. Stands of aspen, oak and white spruce indicate plant life has been established for a long time.

Eventually, the trees give way to the open dunes, accessed by ladders. At the top, think of this: above you a mighty river once flowed, depositing the sand beneath your feet. Since the river shrank to its present size, the sand has been constantly sifted by the wind, creating and recreating the dunes. Excavations have revealed that some dunes may have been buried and resurrected as many as three times over the past millennia.

The expanse of sand often feels like a desert, with a summer surface temperature of up to 55º C, but it's not. A true desert gets less than 25 centimetres of rain a year; here, annual precipitation is almost double that.

At first glance life seems absent, but a closer look reveals the dune's insect residents. Here, thread-waisted, digger or wingless wasps wage small wars in their quest for life; there, tiger beetles or black darling beetles leave their tracks in the sand. Toothed field grasshoppers and rare slant-faced grasshoppers provide food for plains hognose snakes.

Look up. Turkey vultures often circle on

Jerry Kautz

The blue-green Devil's Punch Bowl seems magical, but the explanation for this unusual landform is simple enough; underground streams have eroded and collapsed the sand hills beside the Assiniboine River to create a bowl-shaped depression in the earth.

the wind, searching for carrion (or, as one naturalist quips, waiting for hikers to perish on the trail).

The trail from the dunes to the Devil's Punch Bowl goes through undulating mixed grass prairie, where morning and evening hikers may see white-tailed deer, coyotes or, if you're lucky, a fleeting glimpse of a northern prairie skink, Manitoba's only lizard. Birders should be able to spot clay-colored sparrows, western meadowlarks and least flycatchers along these trails.

Just before the descent to riverside springs, a viewing platform allows an excellent view of the Devil's Punch Bowl, a series of beaver ponds surrounded by white spruce. Beavers have been busy here, damming the creek that runs from the springs, creating an excellent place to spot painted turtles, muskrats or long-tailed weasels. Just beyond, the trail descends to the river, where summer water levels reveal wide gravel and sand bars littered with freshwater mussels.

Plants such as sand dock, inset, grab the sand and form a stable root base for shrubs such as bearberry, creeping juniper and wolf willow. The shrubs are eventually joined by aspen, oak and white spruce, above.

Access The Spirit Sands trailhead is on the west side of PTH 5, north of the river. During the late spring and summer months, mornings and evenings are the best times to hike these trails, for midday sun renders both the dunes and the soft, sandy trails uncomfortably hot. Those not wishing to attempt the hike can book a place on a horse-drawn wagon, operating during summer from a corral near the parking lot.

Especially at dawn or dusk, coyotes, left, are often heard, if not seen.

Southern Manitoba

Epinette Creek Trails

This trail system comprises several loops of varying lengths and a 24-kilometre extension called the Newfoundland Trail. You can take a day hike or experience a week-long backcountry adventure. With diverse terrain, luxurious habitats, spectacular views and wildlife activity, it's not surprising that many of Manitoba's outdoor enthusiasts say Epinette is the best trail system in the province.

Epinette Creek has wound through the recorded history of Spruce Woods since 1785 when Pine Fort (*Fort des Epinettes*) was built at its junction with the Assiniboine River. An important pemmican provisioning post for the North West Company, its success prompted the Hudson's Bay Company to build forts nearby.

The Epinette Trail begins 10 kilometres upstream of the junction, and initially travels through an area burned in a wild fire in 1997. Today, a new forest is taking root. Against a backdrop of blackened timbers, soft green aspen shoots mingle with purple fireweed. White-tailed deer and moose savor the delicate new shoots and are often seen at dusk or dawn.

After leaving the burn area, the trail follows the creek south through white spruce and aspen, crossing it twice. The riparian forest includes black ash, balsam poplar, basswood, elm, cottonwood and Manitoba maple. The ground is often muddy, especially in spring and wet weather.

The allure of the Epinette Trails is the everchanging scenery and terrain. From the four-kilometre Spruce Trail to the 40-kilometre Epinette/Newfoundland circuit, the landscape is constantly varied. You can stroll moist, meandering creek beds, hike steep valley slopes, clamber through dense bush and trees and amble across open grasslands. From a path through a towering stand of spruce, you suddenly come upon a hidden marsh or a tamarack bog.

Habitat diversity means wildlife variety. Even the little northern prairie skink is here, although it might remain incognito, hidden under a tree limb along a sandy ravine. You will likely have more success seeing white-tailed deer, elk and moose; the latter two are especially evident in September, October and November. Coyotes, red foxes, woodchucks and ground squirrels are often seen, as well as beavers along the creek.

Listen carefully and you'll hear krinch-krinch-krinch sounds; look closely and you'll see piles of sawdust at the bottom of each tree. These are signs of an army of sawyer beetles, busy at work on dead tree trunks, left. This combination of dead trees and beetles (and a host of other insects) in turn attracts woodpeckers.

Southern Manitoba

At the trailhead, watch for American goldfinches and red-tailed hawks; along the way, you might see red-breasted nuthatches, golden-crowned kinglets and broad-winged hawks.

Access: The Epinette Trails are the most northerly in the park. Turn off PTH 5 at the trail sign on the east side of the highway; travel one kilometre to the trailhead. Maps provided by the park and available at the trailhead show routes marked for hikers, for cyclists, and in winter, for cross-country skiers. Backcountry camping is offered at five designated sites along trails, including Jackfish Lake at the end of the Newfoundland Trail. Many backpackers claim this is the highlight of the trail. The small lake, surrounded by spruce trees, is at the bottom of a ravine, after a breathtaking view of the Assiniboine Valley from the overlooking slope. A cabin on the lake for winter use can be reserved for overnight stays through the Department of Natural Resources in nearby Carberry.

The eyesight of moose is very poor, but its senses of smell and hearing compensate. Though a frightened moose may crash through the underbrush, these huge animals are also capable of moving almost silently through the forest. It's no surprise that their tracks are twice the size of deer tracks, though they look similar.

Assiniboine River Canoe Route

It's possible to paddle the Assiniboine River from the Shellmouth Dam in west-central Manitoba all the way to Winnipeg, but the 125-kilometre stretch from Brandon to PTH 34 is one of the most interesting sections of this route. This is the waterway used by generations of aboriginal people. Spruce Woods was one of their favorite wintering grounds because of the abundant supply of bison, as well as spruce gum and spruce roots (wattape), materials needed to make birch bark canoes. In the 1730s, it was the Assiniboine River that also brought the first European travellers to the area.

To follow in the wake of these early travellers, you can put in at PR 340 and paddle through the Assiniboine Corridor Wildlife Management Area and Shilo military reserve, past the spot where the North West Company and the Hudson's Bay Company had fur trade posts, to the Stockton ferry where camping is permitted.

Perhaps the most scenic day trip is from the Stockton ferry crossing to the canoe campground just east of PTH 5. Along the way, you can visit the Devil's Punch Bowl from a landing on the river.

Alternatively, you can put in just north of the Spruce Woods campground off PTH 5 and plan a one-day journey to the park's equestrian campground, or continue on for another day to the pullout at PTH 34, north of Holland.

The trip through the park is peaceful. The river is cradled within clean sandy cliffs; dense thickets of aspen, spruce and green ash cloak the banks. Above and beyond the treetops, on the open prairie ridges, grasses mingle with wildflowers.

Along the way there are many opportunities to see white-tailed deer, especially at dawn and dusk, feeding near the river; beavers are abundant all along the river. There are many snapping and painted turtles in this riverine community, as well as ducks and geese.

This is a gentle journey, except in early spring when flows are high. The only drawbacks in summer are sandbars in low water areas, which inspired the old local saying: "If you capsize in the Assiniboine, for God's sake, stand up!"

The portion of the Assiniboine that winds through Spruce Woods is one of the most scenic stretches of the river.

Southern Manitoba

The shells of freshwater mussels litter the banks and bars of the Assiniboine.

Beyond the equestrian campsite, there is a second established campsite on the river, about five kilometres past Big Island. After setting up camp, you can follow a trail to the Hogsback – a high, narrow ridge of sand. The following day, as you continue down the river, look back and you'll see how the ridge earned its name.

At this point you are approaching the remains of the SS *Alpha* – a steamboat that plied the waters between Winnipeg and Fort Ellice near the Saskatchewan border from 1875 until 1885, when it ran aground on a sandbar and was abandoned. If river levels are low, you will be able to see the remains of the old steamboat; for the adventurous, it is even possible to canoe right inside the gaping hull. However, when the river is high, the *Alpha* remains invisible in its watery grave.

One caution about the canoe route: bring lots of drinking water because the river water is undrinkable unless treated.

Just upstream from where PTH 5 crosses the river, a pullout allows paddlers easy access to the Devil's Punch Bowl.

Pembina Valley WMA

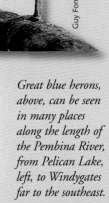

The Pembina River Valley winds more than 250 kilometres through southern Manitoba. In places, it is little more than a narrow stream, elsewhere it broadens into lovely finger lakes and marshes. Almost everywhere, the combination of river and valley creates havens for wildlife. In more than a dozen places, natural areas have been acquired and designated as part of the sprawling Pembina Valley Wildlife Management Area.

The WMA units vary in size, habitat and accessibility, from the Pelican Lake Marsh Unit near Ninette at the valley's northwestern end to the Point Douglas Unit far to the southeast, just north of where the valley enters North Dakota.

The first of these, at the north end of Pelican Lake on PTH 18, is prime habitat for nesting waterfowl. Encompassing 66 hectares of marshes and meadows, it attracts western grebes and a growing number of American white pelicans. From May through July, sedge wrens as well as Nelson's sharp-tailed and Le Conte's sparrows can be found in the wet meadows, while marsh wrens can usually be seen among the cattails.

Downstream, the Grassy Lake Unit east of PTH 34 combines a bountiful wetland with aspen-oak uplands. American white pelicans, double-crested cormorants, western grebes and American coots can be seen in the marshes in spring and summer. In the uplands are Lincoln's sparrows and wild turkeys and, occasionally, a great egret. White-tailed deer winter here in substantial numbers, along

Great blue herons, above, can be seen in many places along the length of the Pembina River, from Pelican Lake, left, to Windygates far to the southeast.

Southern Manitoba

From late summer to early fall, migrating waterfowl stage on wetlands throughout the valley and in October, tundra swans gather in large numbers on the lakes and marshes.

Motorists travelling PR 201 between the Snowflake and Point Douglas Units in late August or early September might come upon a migration of tiger salamanders.

with a variety of fur-bearing mammals, including foxes, coyotes, mink and muskrats.

Farther east, the valley narrows and deepens to become the "Pembina Gorge", as it is locally known. In places the steep slopes are cut by tributary valleys. Thickly treed, they provide prime habitat for deer and moose. In recent years, some people claim to have seen cougars in the area.

The Snowflake Unit follows such a tributary valley east of the town of Snowflake. This is a prime wintering area for deer and includes a sharp-tailed grouse dancing ground, where males may be seen on April mornings performing their energetic mating rituals.

The lower Pembina Valley lies along a major migratory route for hawks. From late March to mid-April, hundreds of red-tailed hawks and several other raptor species may be seen in a single day, along with bald and golden eagles. Paddlers on the river in the spring and early summer may also see mallards, wood ducks, common mergansers and lesser scaup.

In recent years, elk have been seen south of Rock Lake; their bugling calls are sometimes heard in September.

Access This is a sampling of the more accessible riches of the Pembina Valley located about 150 to 200 kilometres southwest of Winnipeg. To fully explore many of the units, tackling the back roads is necessary, or leave the roads behind and hike or paddle into the midst of these special places.

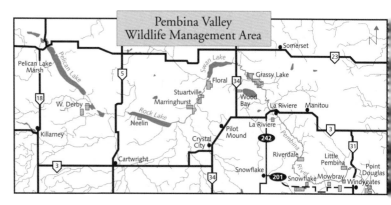

Holmfield WMA

The western meadowlark's bright yellow breast and flutelike song herald spring on the prairies.

Mainly because of its proximity to one of southern Manitoba's main east-west arteries, this small Wildlife Management Area in southwestern Manitoba provides an opportunity to see typical prairie waterfowl "on the fly", as it were.

At 64 hectares, Holmfield WMA is one of Manitoba's smallest. About one-third of it is wetland, the result of an impoundment created by Ducks Unlimited. An area around the marsh was seeded to dense nesting cover and includes reeds and cattails. The balance of the WMA is prairie with bluffs of willow and poplar.

The result is a tiny haven for waterfowl, marsh birds and mammals. Along with mallards, blue-winged teal breed here and Canada geese stage here in April and October.

Red-winged and yellow-headed blackbirds nest in the surrounding reeds and the lilting song of the western meadowlark can be heard in the uplands.

The area also supports a substantial population of white-tailed deer, which are sometimes seen in the WMA, as are foxes and coyotes. Muskrats and weasels, as well as western painted turtles are found in the wetland areas.

Holmfield WMA is located 7.2 kilometres east of the Holmfield turnoff on the north side of PTH 3, about 120 kilometres southeast of Brandon.

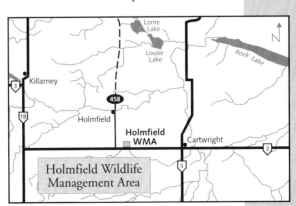

Southern Manitoba

Assiniboine River Corridor - Brandon

The narrow-leaved sunflower is a late bloomer, flowering in August and early September.

Bisected by the Assiniboine River and its deep valley, Brandon has natural advantages for urban wildlife viewing that have been enhanced by developments along the meandering riverbank. The Assiniboine River Corridor combines paved paths and unpaved trails to link existing parks and sports fields along both sides of the Assiniboine with facilities like the Ducks Unlimited (DU) wetlands development just east of 18th Street. The river walk is not just for people – native plants and animals also have a place here.

Prior to European settlement, the Assiniboine was thickly wooded along both banks, with elm, bur oak, green ash, poplar and cottonwood interspersed by small glades and thickets of willow. Many of the elms are gone, but riverbottom forest still exists along the 20 kilometres of trails, providing a home for many species of mammals and birds. The river itself, with its wide meanders, midstream islands and oxbows, is also prime wildlife habitat.

Linda Fairfield

The DU development includes two wetland cells of almost three hectares each, surrounded by dike trails and 1.2 hectares of upland. This varied habitat provides a home for mallards, blue-winged teal and other dabblers. Nesting boxes nearby provide homes for wood ducks and common goldeneye. Nearby, DU's regional office shares space with a regional travel centre.

White-tailed deer can often be seen all along the corridor. Beavers and muskrats find homes along the river. In adjacent woodlands long-tailed weasels and woodchucks may be seen.

Primarily though, this is a place to see birds, especially on the Vireo Trail, just east of the marsh. Accessed by a paved walkway from Kirkcaldy Drive, the Vireo Trail loops through a wooded meander in the river, where red-eyed, warbling, Philadelphia and solitary vireos can be seen, as well as the relatively uncommon yellow-throated vireo.

After visiting the adjacent wetlands, take the pedestrian bridge across the river to the East Dike Trail on the south side of the river. This broad walkway winds along the river through Eleanor Kidd Park, under 18th Street, past a number of playing fields to Queen Elizabeth Park, another good place to see yellow-throated vireos.

During the summer months, a passenger ferry is available to take you across the Assiniboine to the Curran Park Trail, which leads to the park of the same name (with its picnic sites and sports fields) and the west end of the Corridor.

In many places, birders may spot least flycatchers, gray catbirds, yellow warblers, rose-breasted grosbeaks or Baltimore orioles. Curran Park, which is well treed, is home to wood ducks. Spotted sandpipers and willets can be seen along the riverbanks. On the water, pied-billed grebes and American wigeon are common, while hooded mergansers can sometimes be seen.

During fall migration, Swainson's and hermit thrushes stop en route from breeding grounds in the northern forests, as do Harris's sparrows. Tiny Lincoln's sparrows and white-crowned sparrows can also be seen on their journey south.

The white-crowned sparrow, above, joins other southbound migrants that stop along the river in autumn.

Access: The Assiniboine River Corridor can be accessed on the north side from Kirkcaldy Drive, or from 18th or 26th Street on the south side. Parking is available at the regional travel centre, Eleanor Kidd Park, Queen Elizabeth Park and Curran Park, as well as at the eastern trailhead.

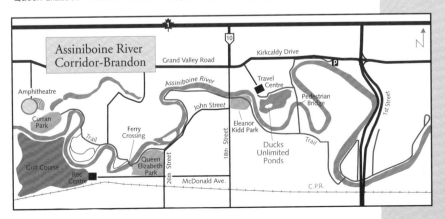

Southern Manitoba

Brandon Hills WMA

Travelling south of Manitoba's second-largest city, the Brandon Hills rise smoky blue on the southern horizon. Part of a long sand and gravel end moraine, the hills form the northwest end of the Darlingford Moraine, which includes the Tiger and Pembina Hills to the southeast.

Running generally east-west, the hills culminate in a steep ridge that rises dramatically to a height of 90 metres above the plains to the east. Undulating and mainly wooded, the uplands combine poplar, bur oak and Manitoba maple forests interspersed with small openings of mixed grass prairie. The 722-hectare Wildlife Management Area was created in the mid-1970s due to the high concentration of deer that the area attracts in winter. The WMA consists of two parcels of land: a large, western block and a narrower eastern block paralleling the edge of the hills.

The well-drained sandy soil boasts diverse vegetation – including beaked and American hazelnut, chokecherry, saskatoon, buffaloberry, highbush cranberry, raspberry and strawberry. Many wildflowers, including pink wintergreen and wild blue flax can be found.

With fluffy white youngsters to feed, this great horned owl will be kept busy hunting small mammals, including voles, mice and even skunks.

Approaching from the east, the Brandon Hills rise steeply from the central plains.

Soon after its establishment, Brandon's cross-country skiers discovered the winter charms of the WMA. Originally developed for the Canada Winter Games, the trail system continues to be groomed by the Brandon Ski Club. The trails also provide excellent hiking and cycling opportunities in the summer. (Motorized vehicles are not allowed in the WMA.)

Among large mammals, white-tailed deer have the WMA largely to themselves, although black bears wander through once in a while. Red foxes, skunks and mink are also found. The Brandon Hills are known for birds, as well as deer. In an inventory undertaken in 1994, 54 species of birds were sighted and of these, 38 were nesting in the WMA. Warblers of many kinds are summer residents, including American redstarts, ovenbirds, yellow warblers and black-and-white warblers. At least three species of vireos breed here, including red-eyed vireos. Rose-breasted grosbeaks can be seen and observant birders may see (or hear) black-billed cuckoos. Listen, too, for the rapid flutelike notes of the veery at dusk.

Bobolinks are occasionally seen, as are indigo buntings. The distinctive song of eastern towhees can be heard in May and June: *drink-your-tea, drink-your-tea*. Nesting boxes along fencelines have aided in increasing the number of both eastern and mountain bluebirds. Tree swallows also use the boxes as nest sites.

The Brandon Hills are among the better places in Manitoba to see Swainson's hawks, above, riding the wind, perched on fenceposts along the road from Brandon, or rarely, resting in a grain field.

Access: Take PTH 10 south from Brandon for 11 kilometres. Turn east, and then south again, following the cross-country skiing signs, to the WMA, a total of 13 kilometres from Brandon. Turn east and follow the road to the parking lot.

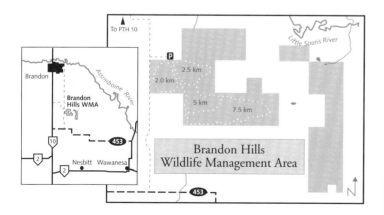

Turtle Mountain Provincial Park

Western painted turtles, for which the park is named, dig nests, lower right, in the gravelly sand along lakes and rivers to lay their eggs.

Like silent echoes of the distant past, a handful of stone tools dating from between 11,000 and 8,000 years ago has been found in Turtle Mountain Provincial Park. These uplands in the southwestern corner of Manitoba were the first to emerge as the glaciers retreated. Here, the earliest Manitobans hunted huge ice age bison and woolly mammoths.

Over succeeding millennia, Turtle Mountain drew a parade of cultures. When Europeans arrived, they too used the ancient upland trails. The Dunseith Trail was used by fur traders as early as the 1800s. Later, in the 1920s and 1930s, rum runners may have used this historic route to smuggle alcohol across the border.

Tucked against the Canada-U.S. border in southwestern Manitoba, the hills of Turtle Mountain rise 180 to 245 metres above the surrounding agricultural patchwork. Turtle Mountain Provincial Park occupies 18,600 hectares of this upland, including at least 200 lakes and one of the largest contiguous tracts of deciduous forest in southwestern Manitoba. A dense understory of shrubs – hazelnut, chokecherry, saskatoon and highbush cranberry – blooms in spring and provides fruit throughout the summer and fall months.

Large numbers of orchids can be found around Mary Lake, along the Dunseith Trail, including calypso and small round-leaved orchids, striped and spotted coralroot and yellow lady's slippers. Eagle Island on Max Lake holds an outstanding example of old oak and elm forest, significant because it is the only area untouched by fire.

The woolly mammoths and giant bison that attracted early hunters are gone, but Turtle Mountain remains an excellent place to view large mammals, including moose and white-tailed deer. Mule deer, which are gone from much of the rest of the province, have also been seen here.

Its black mask and distinctive song — witchity-witchity-witchity-witch — make a common yellowthroat male easy to identify.

Brian Wolitski

Ian Ward

Beaver, usually hard at work, are among the animals a quiet walker might see near Emma Lake.

Winter is a good time to see moose (or signs of moose) along the trails. In the spring white-tailed deer often visit park campgrounds and have been spotted wading in groups along the shores of Mary Lake.

Western painted turtles are abundant – they can be seen along the lakeshores, in wetlands, and sometimes basking in the sun on a park roadway. Nearly every open body of water has a beaver lodge and the shallow rush-lined ponds and lakes also support a variety of other wildlife, such as salamanders, muskrats, raccoons and mink. Birds such as great blue herons, black-crowned night-herons and double-crested cormorants may be seen here and since the mid-1990s there have been occasional sightings of great egrets.

On spring mornings and evenings, ruffed grouse can often be heard drumming, and along the trails, hikers are likely to see a kaleidoscope of avian color. American redstarts, the males resplendent with prominent orange patches on their wings and tails, the females wearing yellow rather than orange, can be found in the forest understory, along with red-eyed vireos and common yellowthroats. The male yellowthroat, distinguished by its black eye mask, bright yellow breast and distinctive call, can be seen among the willows or bulrushes near wetlands.

Early summer evenings here are filled with the laughter of loons and the high, clear song of the white-throated sparrow: *O Sweet Canada, Canada, Canada.*

Red-necked grebes also nest in many of the park's wetlands. To see a large concentration of these strikingly-marked birds, walk to the marsh at end of the Disappearing Lakes Trail in late May or

June. This easy 1.5-kilometre loop off the Oskar Lake road is also a good place to see American goldfinches in their brilliant yellow breeding plumage, and hear eastern phoebes, with their rapid and distinctive call: *fee-be, fee-blee*. Moose (often with calves in the spring) and beaver can also be seen along the trail.

Hikers will share many of the trails with wood frogs, and in the evenings boreal and chorus frogs serenade the coming night.

Spring, summer and fall, hikers on the Adam Lake Self-guiding Wildlife Trail are likely to see moose, particularly in early morning and late evening. This 2.5-kilometre mini-tour of Turtle Mountain's many habitats features a viewing tower overlooking Adam Lake.

In August, non-breeding pelicans can be seen resting on Adam and Max Lakes. Beginning in late September, Max Lake and many of the other lakes serve as resting places for flocks of lesser scaup, also known as bluebills. Feeding in large numbers, they move from lake to lake.

At Gordon Lake, canvasbacks and the slightly smaller redheads can be seen in early to mid-October, feeding on the sago pondweed in the lake. The south side and southwest corner of Nellie Lake also draw these migrating flocks in early October, as does Bredin Lake. After the leaves have fallen, ruffed grouse can be easily observed along many trails.

Eleven trails, from three to 43 kilometres long, offer back-country hikers and cyclists (or in winter, skiers) a range of choices along mostly wide and grassy paths over rolling hills, through towering aspen trees, past many lakes. The James Lake Trail takes hikers or cyclists along 15 kilometres of forested trails and includes a rustic cabin for overnight stays. Riders can take horses on designated equestrian trails, and paddlers can view wildlife

Manitoba's most common lady's slipper orchid, the large yellow lady's slipper, can be seen in June around Mary Lake.

Cow moose often bear young only once every two years, with the number of young largely dependent on the food supply and the severity of winter.

Southern Manitoba

Morning mist greets early risers at the walk-in and canoe campsite on Oskar Lake.

along the two-day, 19-kilometre canoe route that links 10 lakes, including Oskar, Max and James Lakes.

Access Turtle Mountain Provincial Park is situated along the Canada-U.S. border off PTH 10 about 90 kilometres south of Brandon or 15 kilometres south of Boissevain. There are campgrounds at Adam and Max Lakes. Maps are available at the campgrounds and trailheads. Adjacent to the southeast corner of the park is the International Peace Garden, which features walking trails and spectacular summer flower displays.

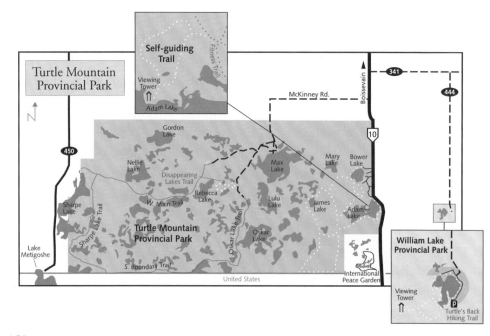

Turtle's Back Trail

Located in William Lake Provincial Park, this five-kilometre trail offers the best viewpoint in the Turtle Mountain region. It owes its 757-metre height to a foundation of ancient uplands that elsewhere eroded away, topped by as much as 150 metres of glacial till.

The trail takes you along the shores of William Lake, up and down hills and through a community pasture to a tower at the summit. The last section is a steep (but worthwhile) climb. From the tower, visitors have a panoramic view over the entire Turtle Mountain area and surrounding farmland from the International Peace Garden to towns along PTH 3.

Many of the animals commonly found in these uplands can be seen here – moose, elk, white-tailed deer, coyotes and red foxes, bobcats, raccoons, beavers and muskrats, as well as many members of the weasel family, including long-tailed, short-tailed and least weasels. Mink, martens, striped skunks and badgers can also be found.

More than 100 species of birds are regularly seen, including many birds of prey – northern harriers and Swainson's, Cooper's, broad-winged, red-tailed and sharp-shinned hawks. The many ponds and lakes harbor pied-billed, red-necked, horned, western and Clark's grebes, as well as at least 12 species of ducks, including redheads, canvasbacks, bufflehead and ring-necked ducks.

Access: William Lake Provincial Park is east of Turtle Mountain Park. Turn east off PTH 10 onto PR 341, then head south on PR 444 to the campground. From the trailhead on the south side of William Lake, it is five kilometres to the summit and back. A second trail around the lake is also about five kilometres, and offers the option of an extension to the viewing tower.

Photographer's Notes

May 25th – "What a spot. It feels like the top of the world. From the tower you have a 360° view, horizon to horizon. Rolling, wooded hills give way to rich farmland. A broad-winged hawk soars over the tree tops at eye level and below, the drumming of a ruffed grouse is drowned out by the laughter of red-necked grebes."

Early spring on the Turtle's Back: within days, the trees will be cloaked in brilliant green.

Southern Manitoba

Souris River Bend WMA

Brian Wolitski

Nesting in woodlands and feeding in open country, the red-tailed hawk is completely at home in Souris River Bend WMA.

Surrounded by gently rolling prairie farmland, the rugged Souris River Bend Wildlife Management Area, 40 kilometres southeast of Brandon, seems almost out of place. The 2,196-hectare WMA is a strikingly beautiful landscape of high cliffs, ravines, terraces and level benchlands with spectacular views and abundant wildlife.

Created in 1968 to provide habitat for white-tailed deer, Souris River Bend contains the largest remnant of native mixed grass prairie in the WMA system. The Souris River makes an abrupt northward turn at The Elbow near the southeast corner of the WMA's main body. This elbow or bend, for which the WMA is named, is the legacy of a water flow reversal in the wake of the last glaciation 11,000 years ago. The portion of the Souris River north of today's bend was once a tributary, flowing south, joining the flow of the Souris eastward toward the Pembina River and glacial Lake Agassiz. Then, as the glaciers retreated and debris blocked the river's former course, the tributary became the main channel and the Souris began to flow northeast to join the Assiniboine River.

The legacy of this dramatic geological history is a dramatic landscape. The southern boundary of the WMA is framed by an imposing southern escarpment along the broad and deep Souris River Valley. North of the elbow the valley narrows considerably but surrenders little of its dramatic depth. The Tiger Hills to the east are picturesque, while Turtle Mountain to the south presents a faint blue outline on the horizon. The slopes of the Souris River Valley, some very nearly vertical, are generally wooded, interspersed with native mixed grass prairie.

The Souris River Bend WMA has a network of hiking and horseback riding trails. Canoeists will also find this stretch of the Souris River challenging, with its sections of white water in spring.

Upland woods cover two-thirds of the WMA, including the valley slopes. On most of the more level areas trembling aspen is the principal tree species, while the slopes are covered by bur oak interspersed with white birch and green ash. The shrub layer is mainly chokecherry and saskatoon bushes with hazelnut and raspberry also conspicuous, all of which is fine for a summertime harvest if you can beat the birds and squirrels. Care should be taken when walking or berry picking, for the ground cover includes poison ivy.

Bur oak is common along the river, but Manitoba maple and American elm lend shade as does the occasional cottonwood.

With the exception of some hayfields producing forage, the open areas are mixed grass prairie. Here, the ground cover contains a wealth of prairie species in seasonal progression from crocus, violet, Indian breadroot, hoary puccoon and pale comandra in the spring through prairie coneflower and fleabane in the summer months to dotted blazingstar, aster and goldenrod as the season wanes into autumn.

As the Wildlife Management Area was created as a haven for white-tailed deer, it's no surprise they can be found in abundance. Coyotes and red foxes are also common residents while elk, moose, and mule deer may be seen on rare occasions. Western painted turtles and large snapping turtles can both be found sunning themselves on rocks or fallen trees by the river.

With a variety of habitats in one area, Souris River Bend is superb for birding. The valley slopes attract orange-crowned and black-and-white warblers, American redstarts and eastern towhees, while the wooded uplands are characterized by Cooper's hawks, black-billed cuckoos, flycatchers, veerys, vireos, yellow warblers, ovenbirds and rose-breasted grosbeaks. Some of these upland birds – the least flycatcher, yellow-throated vireo and yellow warbler – also inhabit the riverbottom forest, along with yellow-bellied sapsuckers. In the grasslands and thickets, characteristic birds include the clay-colored sparrow, mourning dove, eastern kingbird, gray catbird, vesper sparrow, American goldfinch and sharp-tailed

Though peak flows should be avoided, spring is the best time to canoe the Souris River.

Through a curtain of mist over the Souris River, dawn colors both sky and water.

grouse. A red-tailed hawk may circle overhead while the river shallows may play host to a spotted sandpiper or a great blue heron. Canoeists may find belted kingfishers nesting in the steep banks along the river.

From Brandon, travel south about 30 kilometres on PTH 10; take PTH 2 east 7 kilometres to the junction of PR 346. Travel south on PR 346, which forms a portion of the western boundary of the WMA, to one of the designated vehicle trails. For excellent views of The Elbow and the Souris Valley, continue south on 346 until about a kilometre before the bridge. Turn east into the WMA and follow the designated trail across the meadow.

Canoe Access: a two-day journey begins where a municipal road extending directly south from PR 348 crosses the Souris River. It ends at Wawanesa, 58 kilometres downstream. Access can also be had at the bridge on PR 346. While spring is the best time for canoeing (whitewater skills are necessary), peak flows during spring break-up should be avoided. Water levels are usually too low for canoeing after early summer.

Whitewater Lake WMA

Like the Cheshire Cat in *Alice's Adventures in Wonderland*, Whitewater Lake has a habit of appearing and vanishing. A shallow basin with no outlet, dependent upon run-off water from nearby Turtle Mountain for its existence, the lake's water levels have fluctuated considerably over the past century, drying up completely during the drought years of the 1930s and late 1980s. When the water levels are low, the presence of alkali paints the shorelines white, inspiring those who named the lake.

The intermittent drying of the basin is crucial to sustaining the lushness of marsh vegetation, but large numbers of birds are dispersed, often for years at a time. In the mid-1990s, considerable effort was made to offset the effects of drought with the construction of a managed marsh unit at the east end of the lake. Today, thanks to ring-diked water retention cells, there is a dependable haven for large numbers of waterfowl adjacent to Whitewater Lake.

Wildlife watchers have benefitted from the managed marsh as well. Along with a viewing mound and picnic area, a boardwalk and trail have been developed.

Surrounded by nesting cover in the form of cattails, bulrushes and whitetop grass, Whitewater Lake abounds with important duck food such as water milfoil and sago pondweed. Beyond the lakeshore, meadowlands of spikerush and sedge give way to mixed grass prairie.

Huge numbers of snow and Canada geese use the lake as a resting place during spring and fall migrations and the sight of tens of thousands of them lifting into the sky in unison is not easily forgotten.

Brian Wolitski

Preferring shallow water, this northern shoveler pair is right at home at Whitewater Lake.

Photographer's Notes

May 27th: "This is a 'must see' for any birder. I have never seen such a vast number and variety of birds that rely on water. Could have spent three days here if I'd had more film."

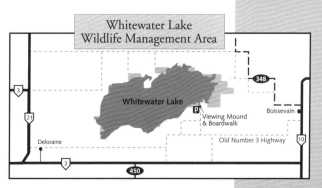

Tundra swans, Canada's most abundant swan, stage here in large numbers in late October.

In spring dabbling ducks – mallards, northern pintails and gadwalls – and diving ducks, such as canvasbacks and redheads, nest around the lake. During May and June, the marsh is full of the chatter of American coots, and horned grebes can often be spotted busily building their nests. Black-crowned night-herons roost during the day, saving themselves for night time fishing expeditions. Lovely American avocets, with their long curved bills, are commonly seen along the shorelines.

In fall, a few greater white-fronted and Ross's geese join the snow and Canada geese. Bald eagles gather around the lakes as the water begins to freeze between late October and mid-November. Freezing temperatures don't send all birds south right away; in late October, one of the province's largest concentrations of tundra swans assembles. They are easily seen with heads held high over graceful necks, feeding on pond weeds in the lake or flying from one spot to another. Adjacent uplands are home to white-tailed deer, sharp-tailed grouse and gray partridge.

Access From Brandon, take PTH 10 south about 75 kilometres to Boissevain. To reach the viewing mound and boardwalk at the east end of Whitewater Lake, turn west at the south end of town onto the gravel road locally known as Old Highway 3, travel 13 kilometres west, then north three kilometres to the viewing area.

Alexander-Griswold Marsh

As its name might suggest, Alexander-Griswold Marsh lies between the villages of Alexander and Griswold, about 35 kilometres west of Brandon. Long and narrow, the 1,050-hectare marsh roughly parallels the south edge of the Trans-Canada Highway for 13 kilometres, but it is best accessed via the municipal roads that crisscross the area east of PTH 21.

Though less well-known than its sister marshes in southwestern Manitoba – Oak Lake-Plum Lakes and Whitewater Lake – Alexander-Griswold Marsh is nonetheless a significant breeding and staging area for waterfowl and shorebirds. The marsh has suffered from drought in past years (notably at its west end, near PTH 21), but Ducks Unlimited has reinforced the marsh basin's dependability by building six water retention cells and securing adjacent upland nesting cover.

American avocets, above, breed in the marsh, and are commonly seen along shallow mudflats.

Canada geese and snow geese use the marsh as a stopping place during spring (mid-April to mid-May) and fall (mid-September to late October) migrations. Dabbling ducks such as mallards, northern pintails, gadwalls, northern shovelers and blue-winged teal nest in the grasslands, while diving ducks such as canvasbacks, redheads and lesser scaup nest in the bulrushes and cattails near the water's edge. Black-crowned night-herons and soras also nest here.

Access A billboard placed by Ducks Unlimited on the south side of the Trans-Canada Highway just west of Alexander (about 25 kilometres west of Brandon) indicates the vicinity of the marsh. The marsh is best viewed, however, along municipal roads south of the Trans-Canada and east of PTH 21.

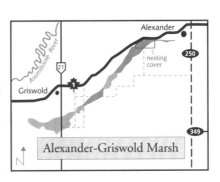

Oak Lake - Plum Lakes

Blue geese arrive in large numbers every spring and fall.

The Oak Lake-Plum Lakes marsh complex in southwestern Manitoba is what remains of a large lake that formed 12,000 years ago in the wake of retreating glaciers. Today, water flows east from Saskatchewan's Moose Mountains into this shallow basin. From there, it slowly drains into the Plum Lakes wetland, a maze of interconnected marshes, and east into the Souris River.

With its wooded ridge separating Oak Lake from the Plum Lakes, this marsh complex embodies one of the best birding areas in Manitoba.

This is an important nesting area for Canada geese and it is used in migration by snow geese. During some autumn migrations, the greater white-fronted geese passing through Manitoba stop at either the Oak Lake-Plum Lakes or Whitewater Lake. The wetlands also draw many species of ducks – ruddy, redhead, canvasback, bufflehead and northern pintail, as well as mallards and blue-winged teal. In fall the area is a major staging area for tundra swans and sandhill cranes.

Oak Lake is also home to large colonies of eared grebes and Franklin's gulls. Some years, the colony of gulls stretches nearly two kilometres along the lakeshore. During the spring, the wooded ridge that separates Oak Lake from the Plum Lake marshes is one of the best places in Manitoba to view a wide variety of birds in a short time. The ridge road leads to a viewing platform overlooking Oak Lake. Beyond it, a 5.6-kilometre hiking path follows the ridge to the Oak Lake dam. Along the way watch

Dense reeds and marsh grass provide the kind of nesting cover that coots prefer, but also suit muskrats perfectly.

for black-crowned night-herons, great blue herons, American bitterns, American avocets, Virginia rails, common snipe and many species of ducks. Western grebes are also common and Clark's grebes can occasionally be seen. Feasting on the riches of the marshes are beavers, muskrats, mink, skunks, foxes and raccoons.

In the forested areas of Oak Lake and the thickets along the shoreline are Baltimore orioles, white-breasted nuthatches, American goldfinches and both eastern and western kingbirds.

Listen for the distinctive song of the Sprague's pipit high above meadow areas. There are a couple of good marshy areas on PR 254 to check for Le Conte's sparrows and Nelson's sharp-tailed sparrows: one is just north of the marina.

The region also provides good viewing opportunities for bald eagles. Peregrine falcons are rare but stop here annually in spring and fall. In winter, check fields and fence posts for snowy owls.

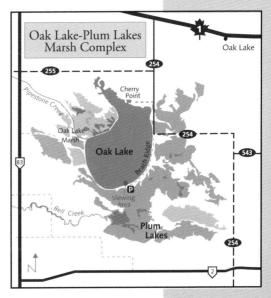

The Oak Lake-Plum Lakes complex is situated about 60 kilometres west of Brandon and 12 kilometres south on PR 254 off the Trans-Canada Highway (Number 1). The access road to the campground, just past the cottage development on PR 254, also leads to the wooded beach ridge between Oak Lake and the Plum Lakes.

Poverty Plains

Jerry Kautz

Tiny and speckled like the pebbles in the surrounding field, the eggs of a vesper sparrow are hidden in plain view.

The sandy soil between Broomhill and Pierson is a souvenir of an ancient beach ridge of glacial Lake Souris that covered a large area west and north of Turtle Mountain 13,000 years ago. Farmers tried to eke out a living cultivating this soil but abandoned their attempts during the drought of the 1930s, leaving the area with a nickname: Poverty Plains.

Today this far corner of southwestern Manitoba is predominantly open pasture land and mixed-grass prairie broken by occasional stands of willow or aspen. This habitat harbors several threatened and endangered grassland bird species including Baird's sparrows, loggerhead shrikes and ferruginous hawks. Indeed, if birding opportunities were the coin of the realm, Poverty Plains would be very much a misnomer.

Backroad travel is key to wildlife viewing in this part of Manitoba, beginning after a 140-kilometre drive southwest of Brandon. A slow trip along gravel roads that cross these grasslands is often rewarded with sightings of prairie specialties such as the upland sandpiper, sharp-tailed grouse, or chestnut-collared longspur. In areas of native grasslands you may hear a grasshopper sparrow or a Sprague's pipit overhead. Walking through the grass may flush a savannah or vesper sparrow during nesting season, and lark buntings during drier periods. This is a thinly populated part of southern Manitoba, where horizons seem infinite under a vaulted sky.

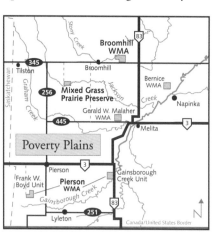

An upland sandpiper surveys its territory, claiming space with a long, slurred whistle.

You can walk undisturbed through the shifting array of grasses and wildflowers, but it's also an area of the province where there is little public land, and permission is required before entering private property. The Broomhill WMA, the Pierson WMA, and the Mixed Grass Prairie Preserve provide opportunities for public access.

Broomhill WMA

Encompassing 324 hectares, Broomhill WMA is largely a mixture of open grassland, shrubs and trees. Native grasses, including spear grass and blue grama are common, but there are many areas dominated by non-native grasses, notably smooth brome grass, as much of this area was once cultivated. Clumps of willow are common and small groves of aspen occur in the north and southeast portions of the WMA. Patches of silverberry, low prairie rose and snowberry are scattered throughout. While no permanent bodies of water exist in the WMA, there is a temporary wetland in the southwest corner and water sometimes collects along the WMA's southern boundary. Much of the WMA has been disturbed in recent years by gravel mining.

Like other parts of the Poverty Plains, Broomhill WMA is a notable habitat for some of Canada's threatened and endangered bird species. It supports one of the largest known concentrations of Baird's sparrows in southwestern Manitoba. This buff-brown ground-nesting bird, distinguished by its distinctive call (two or three zips followed by a musical trill), can best be observed in the WMA's grassy northwest corner. Loggerhead shrikes, another bird on Manitoba's endangered species list, nest among open shrubbery on the eastern portion of the WMA and are often seen perched along the road separating the western quarter-section and the main section. Other frequently observed prairie birds include grasshopper and clay-colored sparrows, western meadowlarks, willow flycatchers, upland sandpipers, Sprague's pipits and chestnut-collared longspurs.

In the spring, sharp-tailed grouse gather on leks (dancing grounds) on or near the WMA to perform their elaborate courtship displays.

Uncommon throughout its range, the Baird's sparrow is perhaps most easily seen here, in the Broomhill WMA.

Access: From Brandon, travel south 24 kilometres on PTH 10, then west 76 kilometres on PTH 2 to the junction of PTH 83. Turn south and travel 18 kilometres to the junction of PR 345. Broomhill WMA is located 3.5 kilometres west on PR 345. Signs mark the area on the north side of the road.

Mixed Grass Prairie Preserve

In 1993, 130 hectares of land were purchased southwest of the village of Broomhill to protect and manage its native prairie. Open to the public, the preserve features a demonstration project illustrating the relationship between grazing animals (cattle in this case, but applicable to bison that once roamed the plains) and plant growth.

Jackson Creek flows through this area, lending a slight roll to the expanse of prairie. Throughout the growing season, visitors will find something to delight the eye: crocuses, delicate crowfoot violets and the reddish-purple three-flowered avens in the early spring, the latter turning in June to "prairie smoke", delicate puffs of pink as the flowers set seed. Blooming in June, the bright blue flowers of blue-eyed grass (really a member of the iris family), complement the golden blooms of yellow stargrass, the only member of the amaryllis family in Manitoba. Cool season grasses, such as needle and thread grass and porcupine grass, shoot up in June while both white and purple prairie clover and yellow coneflowers join the ground cover as the summer moves through July. August sees native sunflowers, goldenrods, common broomweed (with its uncommonly deep yellow petals) and warm season grasses such as little bluestem and blue grama begin to bloom and seed. As autumn beckons, dotted blazing star and asters give the mixed grass prairie the last brush of color.

Birds commonly found in the area include the endangered Baird's sparrow and loggerhead shrike and the threatened ferruginous hawk. Other prairie species, such as the grasshopper sparrow, clay-colored sparrow, chestnut-collared longspur, western

There are many reasons for the decline of the loggerhead shrike. Creatures of habit, these robin-sized predatory songbirds sometimes refuse to breed if their nesting sites have been altered.

Preyed upon by raccoons, crows and some hawks, young shrikes are also regularly killed on highways as they hunt for grasshoppers and other insects. Their numbers have been seriously reduced by pesticides.

The springtime courting ceremonies of sharp-tailed grouse involve lengthy displays of dancing and fluttering, accompanied by deep pigeon-like coos.

meadowlark, upland sandpiper, Sprague's pipit and sharp-tailed grouse may also be seen. The waters of Jackson Creek are home to blue-winged teal, gadwalls, mallards and marbled godwits. Burrowing owls once nested here and artificial nest burrows are evidence of reintroductions that were conducted here for several years during the 1990s.

The tiny burrowing owl, right, is all but extirpated in Manitoba, despite determined efforts at restoration.

Access: From Broomhill, travel west 7.2 kilometres on PR 345, then turn south onto Dublin Road, and travel an additional four kilometres. The preserve is on the east side of the road. A sign at the northwest corner of the preserve marks the spot.

Pierson WMA

Southwest of Melita, very near the corner where Manitoba, Saskatchewan and North Dakota meet, 260 hectares have been set aside to maintain habitat for deer, grouse and grassland birds. The WMA is divided into two distinct units set about 32 kilometres apart: the Frank W. Boyd Unit lies to the west, less than two kilometres from the Saskatchewan border, while the eastern Gainsborough Creek Unit straddles Gainsborough Creek south of Melita along PTH 83.

Uncommon in Manitoba, the ring-necked pheasant can be seen here in the Pierson WMA, as well as farther south. Gainsborough Creek, below, provides habitat for both white-tailed and mule deer.

The east unit is the more beguiling of the two in the Pierson WMA, including as it does a portion of Gainsborough Creek, which provides both topographical relief and treed riparian habitat, a haven for white-tailed deer and the occasional mule deer. The uplands feature native grasses such as spear grass, little bluestem and blue grama, a grass with eyebrowlike tufts.

The west unit contains a mixture of aspen forest and willow interspersed with a number of small temporary wetlands.

 The varied habitat of the Pierson WMA attracts a wide array of birds. These include great-horned and long-eared owls, black-billed cuckoos, red-eyed and warbling vireos and upland sandpipers. Vesper sparrows build tiny nests among the grasses, each with a cluster of spotted eggs, while clay-colored sparrows nest in nearby shrubs. The WMA is also one of only a few known sites in Manitoba for Say's phoebes. One may also see gray partridge, sharp-tailed and ruffed grouse and ring-necked pheasants, also found along PR 256 south of Pierson and PR 251 near Lyleton.

Gainsborough Creek Unit: from Melita, travel south 18 kilometres on PTH 83. This unit is on the west side of the highway, marked with a sign.

Frank W. Boyd Unit: from Pierson, travel 5.6 kilometres west on PTH 3, then turn south onto a gravel road for seven kilometres. A sign on the east side of the road marks the site.

Western Manitoba

Western Manitoba
Prairie Mountains & Parklands

Manitoba's western parkland climbs from the prairie floor in an ascent that is at times almost imperceptible, and at other times dramatic. This is mountain country, Manitoba style, spruce-scented, lake-dotted, home to the highest elevations in the province. This great plateau, defined along its eastern edge by the Manitoba Escarpment, is the province's western backbone.

If you had been standing along the edge of the escarpment during the last glaciation 20,000 years ago, all before you would be ice – an Antarctica of ice. Ten thousand years later, a visitor to the same spot would have seen a vast lake, glacial Lake Agassiz, lapping at the edge of the escarpment's shale caprock.

Today, rising as much as 500 metres above the adjacent Manitoba Lowlands, the plateau is a region of rolling terrain with gentle uplands and broad valleys created by preglacial rivers. The Yellowhead Route (Number 16), follows one of the ancient valleys; PTH 5 runs along one of Lake Agassiz's beaches and PTH 10 north to The Pas is the region's main artery.

Throughout western Manitoba, deep, clear spring-fed lakes, thick stands of spruce and groves of aspen are interspersed with open meadows and occasional bogs and marshes. This diversity gives the region one of the richest and most varied inventories of plants and animals in the province. Little wonder this was the site of

In September and early October, Manitoba's prairie mountains resound with the trumpeting mating calls of bull elk, left.

The Assiniboine River traces a sinuous path through the broad valley cut by its glacial ancestor. This parkland environment provides prime habitat for white-tailed deer and many species of birds.

Western Manitoba

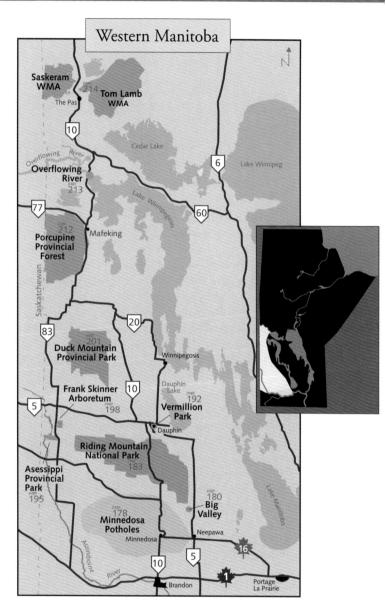

Deep valleys, wooded gorges and long vistas are just some of the delights hikers will find on the escarpment trails in Riding Mountain National Park, opposite.

Manitoba's first national park, Riding Mountain National Park, a sanctuary of protected wilderness about an hour's drive north of Brandon.

Farther north, Duck Mountain Provincial Park encompasses the province's highest point, Baldy Mountain, while the hills of Asessippi Provincial Park, on Manitoba's western edge, sustain mixed grass prairie and aspen parkland at the junction of two river valleys. Here, and elsewhere in Manitoba's western region, are some of the best places in the province to view wildlife.

Trip Planning

If you have a day in the region . . . ❂ Spend it exploring Riding Mountain National Park. Visit the Lake Audy Plains Bison Enclosure, and take a drive along PTH 10 or 19, stopping to hike or bike the trails along the way. ❂ Do some birding at Minnedosa Potholes and spend the afternoon at Big Valley, where children can splash in the creek on a hot summer day.

If you have two or three days . . . ❂ Explore Riding Mountain's western valleys, including the Birdtail Bench, or follow a day in the park with an overnight visit to Dauphin's Vermillion Park. ❂ Take the Yellowhead Highway (Number 16) to the province's western edge and visit Asessippi Provincial Park and Frank Skinner Arboretum.

If you have a week or more . . . ❂ Take a scenic tour of Manitoba's Prairie Mountains, fromTurtle Mountain Provincial Park (covered in the Southern Manitoba section) north up the western side of the province through Riding Mountain National Park, Duck Mountain Provincial Park and Porcupine Hills Provincial Forest. ❂ Spend time in and around Duck Mountain Provincial Park, including visits to the province's high point (Baldy Mountain) and make side trips to Asessippi Provincial Park and Frank Skinner Arboretum. ❂ Fully explore Riding Mountain's backcountry with a combination of driving, hiking, biking or horseback riding. Riding Mountain is also a great winter destination for snowshoeing and skiing, both downhill and cross-country.

Western Manitoba

Minnedosa Potholes
& Proven Lake WMA

Prairie potholes – brimming with water and ringed with vegetation – are so crucial to the continent's breeding waterfowl that they are often collectively called North America's duck factory. Here, a canvasback hen, below right, can raise its young alongside most of North America's other species of ducks.

South of Riding Mountain National Park, the last ice age left a patchwork of small ponds and marshes. These tiny wetlands, called "kettle lakes" or "potholes", were formed by sections of ice that became separated from the larger ice sheet and slowly melted in place. Geographers know this as ground moraine topography, but for ducks this wetland-dotted landscape is high density housing.

For millennia these small marshes, fringed by reeds and cattails, covered with pondweed and sometimes ringed by willows, provided habitat for nesting waterfowl, producing ducks and geese by the tens of thousands. Since settlement, however, these potholes have been drained by the hundreds to increase cropland. Combined with a lengthy and serious drought, these activities brought waterfowl populations to a critical low at the end of the 1980s.

In the past decade, the efforts of conservation agencies and concerned farmers, along with increased precipitation, have combined to return duck populations to levels unseen in many years.

The potholes, round or irregularly shaped and sometimes only a stone's throw wide, draw ducks by the thousands every spring to nest and raise their young. They also support a

178

variety of other birds and mammals, including muskrats, raccoons, red foxes and striped skunks.

More than a dozen species of ducks, as well as Canada geese, nest every year in these small bodies of water. Mallards and pintails are among the earliest spring arrivals in Manitoba, the former returning as early as late March and the latter a week or two later. May and early June are the best times to view the brilliant breeding plumage of ruddy ducks, canvasbacks and horned grebes, and to see mallard ducklings at close range. Throughout the summer, marsh birds, such as yellow-headed blackbirds, marsh wrens, Virginia rails and black-crowned night-herons, can often be seen among the reeds. The surrounding grassy areas are home to Le Conte's sparrows and Nelson's sharp-tailed sparrows.

Just south of Riding Mountain National Park is Proven Lake Wildlife Management Area, one of the larger pothole lakes in the region. This 2,000-hectare wetland is managed with the assistance of Ducks Unlimited to create prime habitat for diving ducks, specifically canvasbacks and redheads, as well as other waterfowl and shore birds. Franklin's gulls nest here, along with black terns. During spring and fall migrations, in mid- to late April and early October, as many as 10,000 ducks, more than 3,000 snow geese, and lesser numbers of coots, gulls, terns and cormorants stage on Proven Lake.

Ducks Unlimited

Nesting tunnels are elevated platforms topped with round wire frames. Placed in a pond and lined with flax and hay, the tunnels give waterfowl, primarily mallards, a safe place to nest, dramatically improving the odds for nesting hens.

Faced with declining duck populations, the Manitoba Habitat Heritage Corporation, Delta Waterfowl Foundation, and landowners began installing nesting tunnels in Manitoba in 1992. By 1997, more than 90 percent of tunnels had nests. Of those, three of four contained successful broods of ducklings. Compared to a normal survival rate of 15 percent for ground nesting ducks, high-rise housing has been an outstanding success story.

> Access
>
> Some of the best areas for viewing potholes and the wildlife they support are just south of Minnedosa, along PTH 24 and PR 250. The ponds are all along both roads and birds are ever-present during the spring, summer and fall. Proven Lake is located just west of PTH 10, less than a kilometre north of the junction with PTH 45.

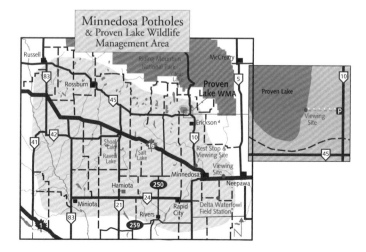

179

Western Manitoba

Big Valley

The American bittern is well camouflaged, particularly in tall reeds.

McFadden Creek, far right, is a magnet for wildlife as it winds through Big Valley.

True to its name, Big Valley is a steep-sided ravine in the Manitoba Escarpment north of Neepawa. Nearly 100 metres deep and 500 metres across, it cradles a clear, spring-fed stream in its leafy depths. The little creek tumbles off the escarpment in a series of waterfalls and rapids, drawing both people and wildlife to the 20-hectare natural area in the lush valley bottom.

The slopes are a sandy shale, thickly treed with trembling aspen, balsam or black poplar, bur oak and birch, interspersed with white spruce. Below, Manitoba maples, willows, saskatoons and raspberries edge the stream and spread out across the floor of the valley to meet the broadleaf forest. The clear running water, thick vegetation and fruit-bearing understory combine to provide a haven for wildlife. The valley, which winds south and east from nearby Riding Mountain National Park, serves as a natural corridor for many species.

White-tailed deer are common, and moose are often seen, particularly where the creek widens into marshes and pools. Beaver and muskrats also live along the stream. Coyotes can be heard on many evenings just before sunset and often in the mornings at dawn. Black bears frequent the area and wolves, though rarely seen, are occasionally heard along the valley at night.

Birds of all kinds are found in the valley. Wrens, warblers and thrushes nest here in the spring, and in May and June the distinctive hollow croak of an American bittern can often be heard. Pileated and hairy woodpeckers, as well as gray jays are commonly seen in the large deciduous trees and conifers. Turkey vultures often drift low over the valley, ever watchful for carrion. Belted kingfishers, which nest in holes they excavate in the steep walls, can sometimes be seen patiently perched on a tree branch above the water, waiting

for an unsuspecting minnow below. In late August and September, migrating sandhill cranes ride the updrafts above the valley.

Big Valley is also known for red-sided garter snakes, which can be seen in spring and fall as they move from their wintering places in caves below the frost line to streams and wetlands for the summer.

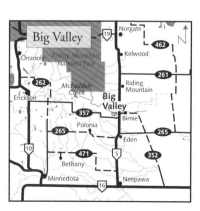

Access: From Neepawa, go north on PTH 5 for 24 kilometres. Turn west at the Big Valley sign on the gravel road and travel for 2.5 kilometres to the road down into the valley. Visitors should use caution or avoid travel on the entrance road in wet or icy conditions.

Balm of Gilead Antiseptic Ointment

Collect one cup of sticky balsam or black poplar buds, below, in early spring. Place in a wide-mouth jar, cover with two inches of olive oil and set in a sunny window. Shake vigorously at least twice a day. After 10 to 14 days, strain, using cheesecloth or old nylons. Squeeze all the oil from the buds.

Heat the oil in a pot and add beeswax to thicken until the cooled ointment is the consistency of hand cream. To test, put some on a teaspoon and stick it in the fridge. If too thick, add more oil; if too runny, add more beeswax. Pour into small jam jars and use on burns, cuts and abrasions.

Western Manitoba

Riding Mountain National Park

Riding Mountain National Park sits astride the province's elevated western spine, which rises dramatically above the surrounding agricultural land. The Riding Mountain escarpment is more than 420 metres above the adjacent Manitoba Lowlands.

Situated where three distinct ecological zones meet, the park's 3,000-square-kilometre landscape provides outstanding diversity. When visiting Riding Mountain, the question is not whether you are going to see wildlife, but how many species you are going to see. Elk, deer, moose and bears are abundant and with some 260 bird species, it is a great place for birding.

The central and most accessible section is characterized by coniferous forests, large lakes and wetlands. The higher western section is aspen parkland – wide valleys filled with rough fescue grasslands, rolling hills and aspen groves. Riding Mountain National Park's eastern section is its most dramatic. Here, a ragged shale cliff forms the Manitoba Escarpment, clearly dividing the Manitoba Lowlands from the higher Saskatchewan Plain. Eleven thousand years ago, the escarpment marked the western shoreline of glacial Lake Agassiz.

White spruce and jack pine crown the crest of the eastern uplands, while hardwood forests of ash, oak, elm and birch cover the face of the escarpment. The moist gorges that cleave the eastern scarp are filled with ferns and mosses.

Many of the park's most charming features, such as this picturesque bandstand near Wasagaming's main beach, were constructed during the 1930s.

Thousands of moose and elk, hundreds of black bears, 40 to 50 wolves, and countless beavers, coyotes and white-tailed deer, as well as more than 50 other mammalian species call the park home. Over half of the more than 260 bird species sighted in the park nest here. Because Riding Mountain lies in the centre of the continent, species from both east and west are seen.

This is one of the best places in the province to see black bears, elk, plains bison, moose, coyotes and beavers. White-tailed deer, foxes and lynx are also here and there are occasional sightings of

Winter poses few problems for plains bison, left, which seem almost oblivious to even the worst weather. Unlike domestic cattle, which sometimes drift with blizzard winds and are crushed against fences and buildings, bison turn their well-protected heads and forequarters into the storm winds and are rarely injured.

Western Manitoba

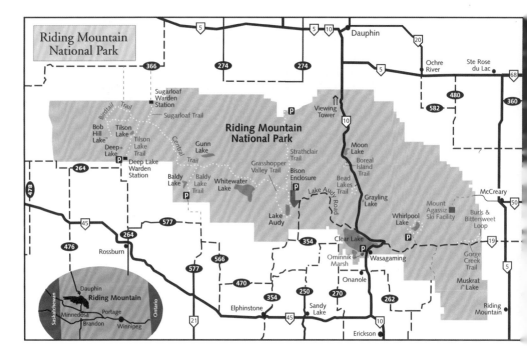

river otters and pine martens. There is an extensive network of day-use and overnight trails designed for hiking, cycling or horseback riding that invites exploration. In many instances, drivers travelling PTH 10 or 19 through the park don't even have to leave their vehicles to see large animals.

Some viewing opportunities are available through interpreter-led outings. On evenings in summer and fall, park interpreters lead car caravans through the park to look for wildlife. In the summer bears are often seen, while elk bugling is offered in September. Coyote and wolf howls are popular events as well.

Riding Mountain park supports Manitoba's largest herd of elk, and the Manitoba subspecies is reported to be the biggest in North America. Its color varies from reddish brown in summer to dark brown in winter, with the rump ranging from ivory to a rusty orange.

Nesting high in conifers, the black-throated green warbler can be identified by its bright golden cheeks, black throat and white wing bars.

Do blonds have more fun? The black bears of Riding Mountain National Park might have an answer. Of the park's estimated 850 bears, nearly half are blond, cinnamon or brown.

Elk may be seen in the Birdtail River Valley in the park's western reaches (see Birdtail Bench on page 191). They also feed along PTH 10 and 19 at dusk and dawn. One of the best times to view them is in September when the velvety fuzz on the adult male's antlers is shed to reveal light gray antlers. Autumn – mid-September to early October – is also prime time for elk bugling.

Because of its wide variety of habitats, Riding Mountain National Park attracts a great diversity of bird life. During spring migration, it is not uncommon for birders to see 100 species in a day. No fewer than 20 species of North American wood warblers have been sighted here, including black-throated green, chestnut-sided, black-and-white, mourning and Canada warblers, as well as American redstarts. The much-sought-after Connecticut warbler can also be found here. The south shore of the Clear Lake Trail, which follows the shoreline of Clear Lake, and the Burls and Bittersweet Trail (see trail description in this section) are both good places to see warblers, beginning in the third week in May and continuing until the third week of June.

The park also draws birders in pursuit of great gray owls, which live year-round near Whirlpool Lake. Spruce and ruffed grouse, boreal chickadees and eight species of the woodpecker family can also be seen. Common loons and red-necked grebes are found on lakes throughout the park. Riding Mountain supports a large raptor population, including broad-winged hawks, merlins, goshawks and sharp-shinned hawks. Bald eagles and ospreys are also seen in the park.

Western Manitoba

Perched motionless atop a post, the great gray owl is easily identified by its yellow eyes and prominent facial discs. Its call is also distinctive, a series of deep whoos, each lower in pitch.

Deciduous woodlands below the escarpment provide homes for species at the northern and western edge of their ranges, such as golden-winged warblers and scarlet tanagers.

Winter activities, including cross-country skiing, snowshoeing and dog sledding, offer their own wildlife viewing opportunities. When the trees lose their leaves, moose and elk are more easily seen. Following an early snowfall, look for bear tracks along the trails, as the park's black bears move to their winter denning sites. Gray jays and white-winged crossbills are often seen, the latter feeding on conifer cones with their scissorlike bills. Hawk owls, pine siskins and evening grosbeaks all inhabit the park in winter.

Access — Riding Mountain National Park is 253 kilometres northwest of Winnipeg. Take the Trans-Canada Highway (Number 1) just past Portage la Prairie, then follow the Yellowhead Highway (Number 16) to Minnedosa and PTH 10 north to the park. A national park pass is required. The resort town of Wasagaming, within walking distance of the park's South Gate, has a Visitor Information Centre, open from mid-May to mid-October. The centre is a good starting place to pick up maps and learn about trails, campgrounds and activities from Parks Canada staff. A bird checklist is available for a small fee.

Pine martens disappeared from Riding Mountain between 1910 and 1920. More than 70 years later, in the early 1990s, 68 martens from neighboring Duck Mountain were released here. Though it has yet to be determined whether the release was successful, pine martens have been recorded at camera stations set up to monitor them.

Wasagaming Area

Located in Wasagaming, the Lakeshore Walk skirts the southeast shoreline of Clear Lake with lovely views out over the water. Beginning at the corner of Columbine Street and Wasagaming Drive, the trail is paved for one kilometre and is wheelchair accessible. Early morning or evening strollers will often see beavers and muskrats, or a great blue heron. Ring-necked ducks, common goldeneye, bufflehead and common mergansers, often with broods of ducklings, shelter in the shallows in May and early June.

Farther south along Wasagaming Drive, at the junction of Boat Cove Road, is Ominnik Marsh with its 1.9-kilometre boardwalk trail, a favorite among children. Ominnik means "pintail" in Anishinaabe, though visitors are more likely to see teal or shovelers. This small lake, created by melting ice when the glaciers retreated, provides attractive habitat for waterfowl, fish, birds, insects and small mammals. From the Visitor Centre beside the park office, you can rent a Marsh Kit that includes a tray, magnifying glass and net to study invertebrates living in the waters of the marsh.

The trail begins with a path through willows and grasses, then moves onto a marsh boardwalk that weaves through reeds and cattails. This is a world where pintails, coots, northern shovelers and blue-winged teal are perfectly at home. Red-winged blackbirds, marsh and sedge wrens and common yellowthroats flitter through the reeds, while black terns cruise overhead in search of minnows.

A northern harrier gliding above suggests the presence of voles and mice, and an evening or early morning visit almost guarantees a glimpse of beavers or muskrats. Mink may also be seen by quiet watchers.

Accessible and absorbing, particularly for young wildlife watchers, Ominnik Marsh is just a short distance from the centre of Wasagaming.

Western Manitoba

Highway 19

PTH 19 is a 34-kilometre gravel road that runs east from Clear Lake through mixed stands of aspen and spruce, then descends the escarpment to the park's elaborate East Gate. Constructed during the 1930s, the gate is now a National Historic Site. At least a dozen day-use trails begin along the highway. At an overlook located just before the road descends the escarpment, there is a spectacular view of the prairies below, land that was once submerged under the waters of glacial Lake Agassiz.

The east trailhead for the 7.8-kilometre Cowan Lake Trail begins across the small bridge near the campground at Whirlpool Lake, on the north side of PTH 19 about 15 kilometres from Clear Lake. Initially traversing an area that is regenerating following a forest fire, the trail provides an excellent place to see moose, particularly in September.

For paddlers, Whirlpool Lake offers excellent opportunities to see wildlife. Bald eagles are often seen here, soaring above the lake. Great gray owls, year-round residents in the nearby forests, can be sighted in the evenings. Three-toed and black-backed woodpeckers may also be seen.

Farther east, the scenic 6.4-kilometre Gorge Creek Trail leads hikers along the escarpment, through lush forest to the edge of a ravine that allows panoramic views across a deep valley carved by meltwater. Wild columbine, hoary puccoon, twining honeysuckle and bur oak line the trail as it descends 320 metres to cross Dead Ox and Gorge Creeks. The trail begins on the north side of PTH 19 and ends at the Birches Picnic Site about six kilometres east. Since it is not a loop trail, it is best to make arrangements to be dropped off and picked up.

Less than a half-kilometre west of the East Gate is a 2.2-kilometre loop called the Burls and Bittersweet Trail. It winds along Dead Ox Creek through a mature deciduous forest where poplars covered with bulbous growths, or burls, abound. Bittersweet is a woody vine that climbs the trunks of trees. Its small greenish flowers produce orange berries in fall.

Both trails provide good bear viewing opportunities, particularly in August and September when they are feeding on acorns and wild berries. Travelling the length of the highway with frequent stops to watch and listen provides one of Manitoba's better opportunities to see and hear songbirds.

A resident of the boreal forest and adjacent parkland edges, the small round-leaved orchid has one- to two-centimetre flowers borne on a tall stalk. Varying in color from rose pink to white, each flower has a prominent lip with tiny rose spots on it. Plants have just one leaf, which gives the flower its name.

Linda Fairfield

Lake Audy Plains Bison

Native prairie – rough fescue grasslands, sprinkled with stands of aspen and spruce forest – is found in the 647-hectare plains bison enclosure near Lake Audy. The cinnamon-colored calves, best seen in May and June, weigh only about 22 kilograms at birth, but within five or six years, bulls can weigh as much as 1,100 kilograms. Mature males are particularly aggressive during the mating season in late July and August.

The Bison and Grasslands Exhibit overlooks the Lake Audy Plain. There you can view the bison, typically in early morning or evening, wallowing or feeding on the wide grassy field. You can drive through the area on your own, armed with the Bison Range Driving Tour Guide (available at the Visitor Centre in Wasagaming), or join special guided tours of the range, which are provided by park staff for a fee.

On the way to Lake Audy and the bison enclosure, there is a marsh on the south side of the road where moose are often seen. At Lake Audy, hikers and mountain bikers can travel the superb 17-kilometre Grasshopper Valley Trail that circles the lake, offering opportunities to see waterfowl, moose, elk and, on occasion, wolves.

Hiking, cycling or horseback riding west on the 67-kilometre Central Trail, or north on the 23-kilometre Strathclair Trail, will also increase your chances of seeing wildlife. Trailheads for both trails are at the parking lot north of the bison enclosure. Especially in September, the first few kilometres of the Strathclair Trail offer some of the best elk viewing in the province. Farther along the trail, look for wolf tracks. The Central Trail leads west toward Whitewater and Gunn Lakes, two more great elk viewing areas, and continues to intersect the Sugarloaf and Tilson Lake Trails, en route to the Birdtail Bench prairie.

Watched by its mother and a brown-headed cowbird, a tawny plains bison calf enjoys its first weeks of life. Born between April and June, bison youngsters resemble domestic calves at birth, but rapidly gain both size and independence.

White admiral and small fritillary butterflies gather on a bison chip near Lake Audy.

 Access — From PTH 10 head north through the park and turn west on Lake Audy Road just north of Clear Lake. Travel 15.2 kilometres to the bison enclosure.

Highway 10

This scenic route, Riding Mountain's most travelled highway, runs north-south through the park's eastern half. A drive along the 53 kilometres from the South Gate entrance to the North Gate may provide you with more wildlife sightings than you've had all year. If you see an animal, slow and stop on the shoulder, but remain in your vehicle. Remember that this is a major highway with heavy traffic at certain times of the year.

Heading north along PTH 10, your first hiking opportunity will be about 28 kilometres north of Wasagaming at the Bead Lakes Trail. This four-kilometre path winds through mature forest, wet meadows, and over a ridge. Beaver dams have affected lake levels here. Keep an eye out for elk, lynx, bear and moose, which are common in this area.

North another seven kilometres along the highway is the Boreal Island Walk, a pleasant one-kilometre ramble through boreal forest vegetation. The route crosses Jackfish Creek to a bog, then passes through stands of black spruce and huge white spruce.

Farther along the highway, a stop at Moon Lake may reveal moose, wolves or their tracks. In early spring, moose are attracted to the highway's edge by the salt used to keep the road free of ice in winter. In July and August, black bears may be feeding near the highway. Birders may find black-throated green warblers, chestnut-sided warblers, golden-crowned kinglets and often a nesting merlin at Moon Lake.

Just before PTH 10 descends into the Valley River Valley, the 12-metre Agassiz Tower offers a breathtaking view of the plains below.

Lingering snow, red willows and a backdrop of dark spruce paint an appealing March landscape along PTH 10.

Birdtail Bench

This picturesque backcountry region in western Riding Mountain is an excellent place to view elk, deer or moose. Particularly between mid-September and mid-October, these large ungulates can often be seen throughout the Birdtail River Valley. Also watch for wolf or bear scat and tracks along the trails. South of where the Tilson Lake Trail joins the Central Trail, there are particularly good views of the valley's eastern slopes.

Birds of prey are very much at home here, drawn by an abundance of small mammals. Red-tailed or occasionally Swainson's and broad-winged hawks can be seen.

Visitors are likely to see antlers along this route or on the hillsides. Do not remove them. Shed by elk and deer every winter, they are an important source of minerals for rodents, such as mice and ground squirrels.

Access This is an area for hikers, cyclists or cross-country skiers. Access is from the Tilson Lake Trail, which begins at the Deep Lake Warden Station parking area off PR 264 on the park's southwestern perimeter. Alternatively, park at the Sugarloaf Warden Station on the northwestern boundary and take the Sugarloaf Trail south for eight kilometres to the Central Trail.

Fast Facts

Cow elk tend to breed with the bull with the largest antlers, strongest odor and loudest, deepest bugle. The harem of a dominant bull can consist of as many as 30 females.

Elk antlers gleam in the sunlight over the Birdtail Bench, above left. The antlers of mature bulls, like the one below, are often damaged during the autumn rut.

Western Manitoba

Vermillion Park

Duckweed carpets the Dead River at Vermillion Park, while nesting platforms provide safe haven for mallard broods, inset.

Tucked away in the heart of Dauphin, Vermillion Park is an emerald haven for breeding waterfowl. The park, which serves as a municipal campground and playground as well as a wildlife sanctuary, is edged on the north and east sides by an old oxbow of the Vermillion River. Dubbed the Dead River by residents, it is in fact anything but, particularly during the months of May and June, when dozens of ducks are fully occupied with the noisy business of raising their families.

The sinuous old watercourse was rejuvenated in 1990 by the City of Dauphin, with assistance from Ducks Unlimited. A variety of nesting platforms, islands and tunnels were installed along the middle of the channel. Vegetation on the riverbank was allowed to grow, creating a margin of cattails, grasses and shrubs. On the opposite bank, homeowners followed suit, developing a series of beautifully landscaped buffer zones with ferns, tall flowers and overhanging plants.

The result is one of the most attractive and accessible places in the province to view ducklings at close range, as well as other creatures associated with a wetland environment.

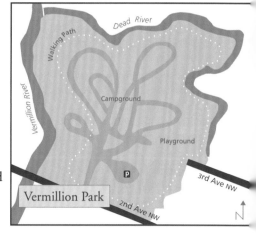

Visitors to the park between May and July can count on seeing at least four species of ducks: mallards and common goldeneye, both commonly found on Manitoba ponds and lakes; hooded

192

Divers & Dabblers

Ducks are categorized in two main groups: divers and dabblers. Diving ducks, such as the hooded merganser, left, as well as lesser scaup, common goldeneye, bufflehead and ring-necked ducks, among others, feed on aquatic plants and invertebrates by diving in deep water.

The colorful wood duck, below, is a dabbler, but also has sharp claws that allow it to perch in trees. Adopted as the mascot for the 1999 Pan American Games, the wood duck is one of North America's most beautiful ducks.

Brian Wolitski

mergansers, the male with his striking black and white crest, and wood ducks. The females, trailed by up to a dozen youngsters, can be seen tracing a pattern of trails through the lacy green coverlet of duckweed, or dozing on a nesting platform next to a fluffy pile of ducklings.

Nesting boxes provide homes for wood ducks, common goldeneye and mergansers, but some prefer the large elms and willows that line the oxbow and the Vermillion River, which flows past the park on the west side.

The lovely old trees draw other birds, including eastern screech-owls as well as downy and pileated woodpeckers. The last of these carve large cavities that are used by the tree-nesting ducks and tree swallows. Red and gray squirrels are also found in abundance. Muskrats are found in both the oxbow and the river, and western painted turtles can often be seen sunning on logs or along the banks.

 Access — From the south follow PTH 5A/10A into Dauphin. Travel north and turn left or west onto 2nd Avenue NW. The entrance to Vermillion Park is just east of the bridge over the Vermillion River, on the north side of the road.

193

Western Manitoba

Asessippi Provincial Park

This is the meeting place of the Assiniboine and Shell River Valleys, "underfit" or "misfit" descendents of the torrential waters that carved these broad, steep valleys.

Lake of the Prairies, 67 kilometres long, was created in 1964 by the construction of the Shellmouth Dam to control flooding along the Assiniboine River. It is one of Manitoba's premier sites for walleye angling.

Mixed grass prairie meets aspen forest here, creating a feeling of spaciousness typical of parkland habitat. Wide vistas of grasses – little bluestem, June grass and blue grama – create rolling meadowlands that seem to overflow from early spring to late fall with ever-changing bouquets of wildflowers. April's crocuses are followed by June's hoary puccoon and western red lilies. July's western wild bergamot is succeeded by goldenrod in August.

Red-osier dogwood is common here, along with wolf willow, with its distinctive pungent fragrance when in bloom, and hawthorn, buffaloberry, pincherry, saskatoon, wild raspberry and chokecherry. The wooded areas consist mainly of aspen and balsam poplar, bur oak and birch. In wetter areas expect Manitoba maple and willows.

White-tailed deer travel the deep valley bottoms, following well-worn trails. Black bears are seen more frequently than in the past and Asessippi is also home to coyotes, raccoons, beavers, skunks, porcupines and a multitude of Richardson's ground squirrels. Moose and elk are seen on occasion.

In open areas look for horned larks, American goldfinches, common nighthawks and kingbirds. Overhead watch for turkey vultures. Ruffed grouse, rose-breasted grosbeaks, warblers and vireos can be found

In April, rough-legged hawks, left, are among nearly a dozen species of migrating hawks that can be seen at Asessippi.

Nearly invisible in the long grass, a white-tailed doe is alert even while resting.

Western Manitoba

Ermine – weasels in their white winter coats – depend on keen hearing and an acute sense of smell to prey on mice, voles and even rabbits. Their tracks are similar to mink tracks, but the straddle and the track size are considerably smaller.

in wooded areas. Yellow-throated vireos, uncommon throughout Manitoba, nest here at the northwestern edge of their range. Red-tailed hawks and, more rarely, Swainson's hawks can be seen flying over the valley in search of prey.

Lake of the Prairies attracts American white pelicans, several species of ducks, great blue herons and belted kingfishers. The lake is also an important staging area for western grebes during migration.

 Access: Asessippi is located on PR 482 west of PTH 83, about halfway between Russell and Roblin and about 220 kilometres northwest of Brandon.

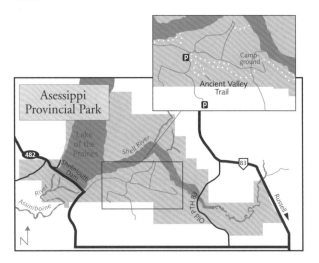

Ancient Valley Trail

This physically challenging trail is one of Asessippi's most beautiful. The three-kilometre self-guiding loop winds along the slopes of the Assiniboine River Valley, allowing hikers an opportunity to appreciate glacial features created more than 9,000 years ago, and see the erosion that continues to shape the valley today.

Habitats along this trail are similar to those found in the rest of the park. Wooded slopes are dominated by trembling aspen and balsam poplar, while the underbrush is thick with shrubs full of edibles – saskatoons, gooseberries, chokecherries, wild raspberries, strawberries and highbush cranberries. Keep an eye out for hawthorn bushes, readily identified by their finger-length spikes. They are painful if you get caught in them. Throughout are open areas of mixed grass prairie dotted with wildflowers.

White-tailed deer, black bears, coyotes and smaller mammals such as rabbits, chipmunks and squirrels can all be seen along this trail. There are benches en route where you can catch your breath and enjoy many of the birds common to the park, including red-tailed hawks and turkey vultures.

In September, sandhill cranes can often be seen overhead, moving south in leisurely flocks of up to 50 or more, their passage accompanied by their characteristic loud rattling call. Whooping cranes, among the rarest birds in the world, have also been sighted here, as the park lies along their narrow migratory corridor to Texas. They are easily recognized by their long necks and legs, black-tipped wings and vibrant, trumpeting calls.

Prairie roses scent the valley air beginning in late June. The fruit, or hips, that follow are eaten by deer, bears, squirrels, grouse and thrushes, among others.

Lake of the Prairies, which fills the Assiniboine Valley here at Asessippi, stretches north and west 67 kilometres, extending into Saskatchewan.

Western Manitoba

Frank Skinner Arboretum

This young pileated woodpecker, right, will grow to 48 centimetres in length, nearly three times the size of a downy woodpecker.

A stroll through this wooded glen demands more than a casual glance. Each labelled tree and plant draws you into the remarkable world of Frank Skinner, pioneer horticulturalist, who, over decades, gathered and planted seeds from around the world and nurtured them to full maturity, despite a harsh prairie climate.

Today, this Provincial Historic Site contains nearly 250 plant species introduced to Manitoba, and features the oldest, most extensive collection of trees and shrubs found on the Canadian prairies.

The rolling hills and deep valleys of Manitoba's parkland provide a picturesque backdrop for Skinner's exotic trees. Among them are native species such as aspen, black poplar, peachleaf willow, Manitoba maple and red-osier dogwood. Equally at home are Manitoba varieties plucked from across the province: white spruce, as well as American elm, green ash and American basswood, usually seen growing in the Red River Valley, and eastern white cedar from eastern Manitoba.

Also impressive are the red pines, usually more at home in Manitoba's southeast corner. Among exotic trees, majestic Scots pines top the list. Planted between 1911 and 1925, these were Skinner's first successful attempts with European seeds. They make a dramatic showing with their orange peeling trunks.

The more unusual trees here include Manchurian pears and Japanese tree lilacs, native to Manchuria and Siberia. There's a Swiss stone pine – the oldest in the province – and also, far removed from its west coast habitat – a Douglas fir. The Finnish spruce thrives here as does the Siberian larch – more at home in western Asia. Look for the Mongolian linden – the only specimen in Manitoba – and the Tibetan purple cone spruce, native, naturally, to Tibet.

There are three self-guiding trails, two about an hour long and the shorter Wild Willow Trail.

Dolores Haggarty

🔭 The arboretum is remarkable not only for its bountiful flora, but also for the abundance and diversity of its wildlife, attracted by an ever-changing menu of seeds and fruits.

White-tailed deer are here, as are black bears. In the fall the latter are often sighted ambling through the arboretum, feasting on mature fruit. The heavily wooded area also supports other mammals such as coyotes, foxes, raccoons, least weasels and mink. Muskrats and an occasional beaver can be found in the arboretum's two wetland areas. Lynx reside here, but like wolves, only their tracks betray their presence.

The bird list is long and varied, from rare and interesting species to more common ones. There have been reported sightings of a scarlet tanager, well out of its range, and more recently, an indigo bunting. The much-sought-after Connecticut warbler can be seen here fairly regularly. Red-eyed vireos, American redstarts and goldfinches are commonly seen, as are great crested flycatchers. Yellow-rumped warblers are frequent nesters.

Other visitors include American avocets, marbled godwits, lesser yellowlegs and cormorants that drop in from nearby wetlands. American white pelicans from Lake of the Prairies often delight visitors with a feathered flypast.

During fall migration in September and early October, a variety of warblers and sparrows stop over, as do juncos, waxwings and purple finches. Pine and evening grosbeaks stay on during winter, feeding on maple and lilac seeds.

Access The Frank Skinner Arboretum is located 31 kilometres north of Russell on PTH 83. From the junction of PR 366, just west of Inglis, continue north 13 kilometres to trail signs on the east side of the highway. Turn east for one kilometre to the arboretum entrance.

Frank Skinner

Frank Skinner arrived with his parents from Scotland in 1895. Though just 13, he went to work raising cattle, riding the open range. Long days on horseback gave him an opportunity to study native plants. Soon, he was transplanting seedlings to the Skinner's garden, and in 1911 he began to introduce plants from Europe.

Those early experiments were mostly unsuccessful and thus began his lifelong quest to develop exotic seeds hardy enough to thrive on the Canadian prairie. Most of the seeds he obtained by mail, through his extensive correspondence with botanists and horticulturists all around the world. Frank Skinner's passion continued until his death in 1967; his work is continued by his son, Hugh.

Western Manitoba

Duck Mountain Provincial Park

Duck Mountain forms the middle (and loftiest) section of Manitoba's western uplands. Duck Mountain Provincial Park, situated northwest of Dauphin includes 1,423 square kilometres of this high plateau.

This is a special place, where water and ice have combined to construct, rather than erode. About 12,000 years ago, as the continental ice sheet retreated northeastward, the great glaciers stalled over Duck Mountain. Perhaps its elevation slowed the retreat; whatever the reason, the ice was kept in place long enough to leave a thick crown of glacial debris, more sand and silt than blanketed other parts of Manitoba's highlands. In places, the glacial till covering "the Ducks" is as much as 250 metres deep, and this topping contributed to its distinction as the highest point in the province (see Baldy Mountain on page 204).

When at last the ice disappeared, life renewed its claim and the peaks and valleys of the raw, reborn landscape eventually supported a mixed forest with abundant wildlife. Boreal forest meets aspen parkland at Duck Mountain. The high rolling hills are topped with white spruce, jack pine and balsam fir. These majestic conifers flow down the flanks of the escarpment where they first intermingle and eventually give way to leafy stands of aspen, poplar, birch, willow and oak.

Grasses and wildflowers blanket the forest openings, edged with chokecherry, saskatoon and snowberry shrubs. Wild mint and wild strawberries grow in abundance, creating layers of color and scent.

The park is home to large numbers of elk, moose and white-tailed deer. Black bears, red foxes and coyotes are all regularly seen. The more remote areas of the Ducks are also home to martens, fishers, mink and wolves, which though rarely seen, are sometimes heard at night.

Found in migration on forest edges adjacent to meadows, the white-crowned sparrow, above, is easily identified by its bold black-and-white crown

Duck Mountain is full of special places, but the Shining Stone Trail at West Blue Lake, far left, is one of the most appealing.

Western Manitoba

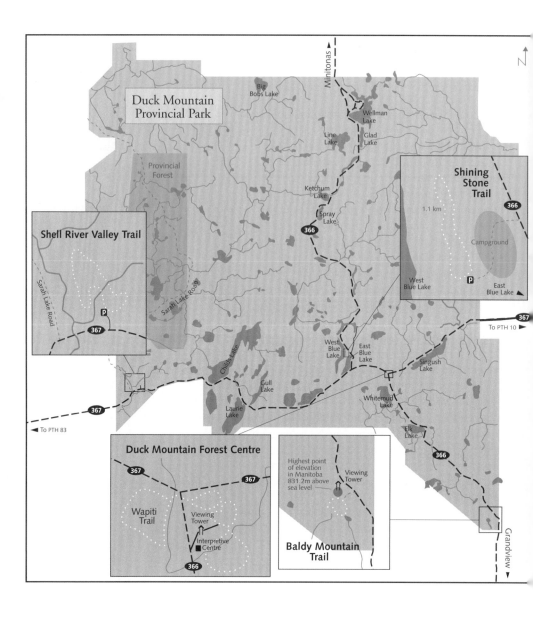

No description of Duck Mountain would be complete without mention of trout fishing. For an ever-increasing number of fly fishing aficionados or simply for Sunday afternoon, string-on-a-stick types, there are enough lakes and fish to suit everyone. Besides lake trout that thrive in the deep glacial lakes, there are stocked splake as well as Arctic char, rainbow, brook and brown trout. Walleye, northern pike, perch and muskellunge may also be found in some lakes.

Duck Mountain is an acclaimed site for bird watching. Its diverse habitats attract a wide range of species. Forests and forest openings are alive with sparrows and warblers, including the rare golden-winged warbler. The American woodcock is also sometimes seen. The park's lakes and wetlands attract waterfowl, including common goldeneye, common mergansers, bufflehead and lesser scaup, as well as pied-billed and red-necked grebes.

Birdwatchers may also focus their binoculars on bald eagles, great blue herons, common loons, turkey vultures and pileated woodpeckers. Duck Mountain is also a superb location for butterfly watching in spring and summer.

In the winter, Duck Mountain is a good place to see evening grosbeaks and crossbills.

Access From Dauphin, follow PTH 5 west to Grandview and turn north on PR 366. Turn east on PR 367 to get to Singush Lake or turn west on PR 367 to access Childs Lake. (PR 367 crosses the park's midsection from east to west.) Continue north on PR 366 to reach the Blue Lakes or Wellman Lake.

Black bears are perfectly at home in the woods. As youngsters, they can shoot up a tree in seconds and as adults, they sometimes claw trunks, perhaps to display their height to rivals. As this big male demonstrates, trees also serve as great back scratchers.

203

Baldy Mountain

One longtime resident of the area describes the Baldy Mountain Trail as "steep and stony", a refreshing departure from the conventional view of Manitoba as flat and featureless. Not only is Baldy Mountain Manitoba's highest point at 831.2 metres, but the trail winds among outcroppings of Cretaceous shales and is sprinkled with glacial erratics, reminders of Manitoba's ancient past.

Visitors may drive to the top of the mountain, located at the southeast corner of the park, climb the wooden lookout tower and enjoy a spectacular 360-degree view of forests and distant farms. This is also an excellent place to view songbirds from tree-top level. Then don hiking boots, search out a solid walking stick and, from the bottom of the tower, head onto the trail.

The 2.5-kilometre loop, blazed and maintained by generations of loggers, forest rangers, fire guardians and game wardens, descends first into a forest of mature spruce and birch. The path here is as promised – steep and stony – but soon reaches an old cabin and stables constructed of enormous logs. They remind us of the early 1900s, a time when people poured into this part of Manitoba on the newly completed Canadian Northern Railway to farm and log the thick forests.

Today the abandoned cabin site is a favorite spot for moose and white-tailed deer. A quiet approach may find them nibbling the soft green undergrowth that surrounds the area. At the very least, there are always plenty of fresh deer and moose tracks and, in the spring, many young trees and shrubs sport the winter pruning effects of ungulates.

From here the trail continues southwest, descending again into deep forest, over small bushy hillocks and around low wet bogs. Hiking the trail in late July, you might spot the blue flower spikes of Kalm's lobelia or the fragrant, waxy blossom of one-flowered wintergreen. Eventually you'll reach a small lake, overflowing from a large beaver pond. Here, at the halfway point, the trail leads out

Found in sunny spots along moist forest edges, the small-flowered columbine, above right, can be distinguished from its wild columbine cousin by its blue and cream flowers. Each long, drooping stalk bears but one blossom, which may appear from late June to early July.

Linda Fairfield

to a viewpoint above a wetland, where you can see waterfowl and beaver activity.

The return is a strenuous climb back to the cabin and beyond it, to the trailhead.

Access Baldy Mountain and the viewing tower are on the west side of PR 366 in the southeastern corner of the park. A short side road leads to the parking area, from which the tower can be seen. The trail begins at its base.

The view from Manitoba's highest point is enhanced by a viewing tower. From the top, particularly in early summer, Duck Mountain's forests seem a tapestry of green as they roll away in every direction.

Duck Mountain Forest Centre

The Duck Mountain Forest Centre was established by the Manitoba Forestry Association to provide information on the history of the area, as well as recreational and picnic opportunities for families and groups.

Irregular in occurence throughout their range, red crossbills, such as the female at right, can be found in spruce forests year-round.

Displays inside the Frank Marvin Visitor Centre recount the history of the logging industry from its earliest years, offer useful information about local plants and animals and provide an overview of the region's early history. Reproductions of native artifacts serve as a reminder that this region was used for thousands of years by the first Manitobans.

Visitors can choose one of the two walking trails that begin at the centre. Both are labelled "easy to moderate". The 2.2-kilometre Pine Meadow Self-guiding Trail loops north of the visitor centre through diverse forest habitat and leads into one of the oldest jack pine forests in the province. Along the way, Blackburnian warblers can sometimes be seen. The males are known as "firethroats" and can be easily distinguished by their bright orange head and throat markings. The same species, with its high, thin song, can often be heard in the park's campgrounds.

Striped coralroot brightens coniferous forests in early to mid-June with its crimson-on-cream flowers.

The Valley River Trail is slightly longer (2.5 kilometres) and travels through meadows and open grasslands, crossing the Valley River twice. There's some climbing involved here, but it's well worth the effort. Near the beginning of the trail, on a high mound, an observation tower provides yet another spectacular view of the Ducks. (Near this spot, at the base of the tower, you can veer off the trail to see an early tipi site.)

In marshy areas, beaver activity is evident. The trail also provides an excellent opportunity to hear, if not see, pileated woodpeckers. Duck Mountain has a high concentration of Manitoba's largest woodpecker, and the bird's slow, rhythmic hammering can be mistaken for someone felling a tree.

Access: The Forest Centre is located just south of the junction of PR 366 and 377. It is open in July and August, but trails and even the access road may flood following periods of heavy rain.

Wapiti Trail

The Wapiti Trail, located near the Forest Centre, shows visitors an example of wildlife habitat management. Here, the Department of Natural Resources used a controlled burn to improve habitat for elk, also known as wapiti. Wapiti is the proto-Algonquian name for elk, and means "white rump". Duck Mountain has the second-largest elk population in Manitoba (Riding Mountain National Park's herd is the largest).

His antlers cloaked in velvet, this bull elk is a portrait of magnificence.

The Wapiti Trail can be a challenge in wet conditions. Portions of the trail are susceptible to flooding from spring runoff and rain. Visitors should check with park officials for trail conditions and wear waterproof boots. Return distance for the long loop is 4.5 kilometres, while the distance to the viewing area and back is about 2.3 kilometres.

The best times to view elk are at dawn and dusk. Visitors may try to call elk with one of the many commercially available calls, but be cautious; these animals may look as harmless as domestic cows, but they are wild and can become aggressive, especially in early fall during the mating season.

 The trail is located just south of the junction of PR 366 and 367.

Shining Stone Trail

If ever a trail invited walking, it must be the Shining Stone Trail. Locals describe it as "short and sweet, beautiful, tranquil and serene".

This nearly-perfect trail (just over a kilometre long) meanders around a small peninsula that juts out high over the northern end of West Blue Lake. At first glance it may seem nothing more than a bucolic after-lunch stroll, but like the one at Baldy Mountain, this trail leads directly back to the last ice age.

Some 12,000 years ago, the Laurentide ice sheet stalled here for a time in its retreat to the northeast. When it moved on, it left behind a deposit of gravel, clay and boulders in the form of a giant finger, looming 30 metres above the lake.

Jerry Kautz

In autumn, bunchberry leaves are a bright reminder of a summer past. In spring, deer browse on the new leaves. Later, many birds including grouse, sparrows, thrushes and vireos, seek its berries.

The little peninsula would have been completely eroded by weather and water, but for vegetation that quickly took hold. The trees, shrubs and undergrowth have kept the lake from eroding the shoreline and prevented soil from sifting into the lake. Even today West Blue Lake remains as clear as glass. When you peer into its depths, the rocks below appear to shine, inspiring the name "Shining Stone".

The trail begins with a short boardwalk over a shallow wetland. High water in the past few years has not only necessitated the boardwalk, but engulfed a margin of shoreline forest that edges the peninsula, slowly drowning it. Deeper water presents no obstacles to the resident common loons, however. This is one of the better places in Manitoba to view them at close quarters, for they can often be seen on the narrow bay adjacent to the trailhead, or on the lake itself. At night, their calls echo through the adjacent campground – the quintessential sound of the Canadian wilderness.

Using a self-guiding trail brochure, visitors are led through three distinct habitats: a natural root staircase takes you to the top of the crest where, surrounded by mature white spruce and jack pine, you'll enjoy a wonderful view of the deep, clear lake. (All is silent here because motorboats are prohibited on West Blue Lake.)

In addition to loons, common mergansers, with the dark green-headed male and his auburn-coiffed partner, can sometimes be seen.

Heading down the steep slopes, dogwood, snowberry and white birch share the ground with maturing white spruce, and at the bottom of the peninsula, even younger white spruce compete for space with red elderberry and birch saplings.

Small forest mammals and birds, including dark-eyed juncos, are plentiful, and share the abundant vegetation for shade, protection and food.

In other areas of the Ducks, logging and fires are the main instruments of forest succession. Here on the Shining Stone Trail, it is windstorms, which blow in off the lakes, that cause changes. The old, weak trees topple and die. Sunlight streams into the new forest window and new growth, such as hazelnut and wild rose, take hold quickly, providing food for a new generation of forest animals and birds.

Fast Facts

Fish stay active during the winter, though their activity levels are reduced. This allows them to conserve energy and enables them to survive colder temperatures. Water below the ice remains at a constant 4° Celsius throughout the winter.

Access The trailhead at West Blue Lake is located on the west side of PR 366, less than two kilometres north of the junction with PR 367. It is just west of the main beach area and campground, adjacent to the boat launch. For those wishing a longer outing, the Blue Lakes Trail extends from the southern end of the Shining Stone Trail.

Deep and clear, West Blue Lake, left, is one of the best places to see common loons in Manitoba.

Shell River Valley Trail

Situated in the southwest corner of the Ducks, this trail offers diverse habitat and terrain – steep ravines, sandy shorelines and a calcium bog – and vegetation ranging from chokecherries and highbush cranberries to ostrich ferns and orchids.

At three kilometres, the trail isn't long, but what it lacks in distance it makes up in altitude. It takes sturdy legs to climb the east ravine wall to the top of the escarpment, but the reward is a breathtaking view of the Shell River Valley, once a glacial spillway and now a multi-hued patchwork of trees, wetlands and grasslands.

Besides a wide range of habitats, the valley is home to many species of animals. For thousands of years this was a favored place for hunters. Archaeologists have discovered a Clovis spear point, as well as mastodon bones, meaning early Manitobans were living and hunting here nearly 11,000 years ago. Centuries later, about 1900 as the story goes, a trapper arrived in the Shell River Valley with two empty wooden wagons. Weeks later, he left with the wagons overflowing with lynx, fisher, cougar and wolf pelts. It is believed he returned to trap each autumn for several years.

Today's visitors are almost as likely to see a cougar as they are a mastodon, but sighting elk and white-tailed deer in open areas during the summer, or moose seeking shelter from northern winds during the winter, is quite conceivable. Pine martens slip through the bush and there are black bears, foxes and badgers. Along the river and creeks, beavers, as well as otters and muskrats are often seen.

The trailhead begins just north of PR 367, heads east to a small bridge over a creek and begins to climb. The grade is steep, but from the top you may be able to see elk in the openings along the Shell River Valley below.

Long a symbol of Canada, beavers create habitat that benefits many other creatures. In Duck Mountain, these include ring-necked ducks, common mergansers, blue-winged teal and horned grebes.

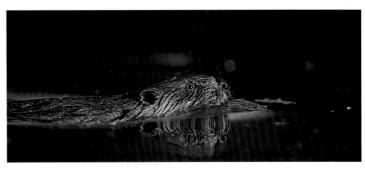

Wetlands, like this one in the Shell Valley, provide perfect habitat for moose and support many types of orchids.

The trail now descends northwest into the valley and the habitat changes, and changes again, from aspen forest to dark evergreens and back to deciduous forest. The trail then crosses an opening in the forest, passing a huge slash pile of slabs and edges from an old sawmill. Ostrich ferns and highbush cranberries around the pile are an indication that the mill closed many years ago. Today, the area belongs to moose and white-tailed deer.

Still heading northwest, the trail fords a larger creek, passes through heavy hazelbush and mixed poplar and crosses a bog on a well-built boardwalk, allowing a close view of an array of vegetation. Besides spongy sphagnum moss and shrubby Labrador tea, there is bog laurel and, thriving in this dark place, delicately colored orchids – striped coralroot, showy lady's slipper, ram's head lady's slipper and hooded ladies'-tresses.

Across a bright opening and swinging south, a spur trail takes you to a beach on the Shell River. Then back to the main trail, through meadows and forest until you come upon another bog. Unlike the first, which was filled with black spongy soil, this one is filled with a white clay substance called calcium carbonate. Rock-hard to the touch, it gives the habitat its name – a calcium bog. The trail passes through mixed willow and replanted white spruce to the trailhead.

This route is a birder's delight. Songbirds fill the forests while the tall grasses are home to common snipe. The wetlands harbor the seldom seen but often heard sora. Great gray, horned, barred and boreal owls are also residents here. Overhead may be red-tailed hawks and Swainson's hawks, or majestic bald eagles.

Pileated and other woodpeckers nest in the mature forests, searching out insects and beetles in rotting trees; grouse – both spruce and ruffed – are in abundance.

 The Shell River Valley Trail is about eight kilometres west of Childs Lake, on the north side of PR 367.

Porcupine Provincial Forest

Porcupine Provincial Forest, a region of forested uplands and clear lakes, lies along the Saskatchewan border north of Duck Mountain. Like Nopiming Provincial Park in eastern Manitoba, this is a good place to view forest succession following a fire.

The blaze began across the border in Saskatchewan in May 1980. By the time it was brought under control, the southwestern and central parts of the hills were extensively burned. Today, these areas nurture a forest of young jack pine and spruce on the uplands and a mixed woodland of aspen, birch and an understory of juvenile jack pine on the slopes. Areas that were not burned typically harbor a mature forest, mainly spruce and birch that grew following a forest fire in the early years of the 20th century.

A combination of habitats makes Porcupine Provincial Forest a haven for large mammals.

The 1980 fire devastated the forest and its inhabitants, but the results of the blaze have been mainly positive, creating large areas of browse for moose and white-tailed deer. A wooded fringe around North Steeprock Lake provides ideal habitat for many smaller mammals, including pine marten and mink.

Many woodland bird species, including Connecticut warblers, can be seen here. Great gray owls and spruce grouse, as well as black-backed and three-toed woodpeckers may be seen in mature forest. Flocks of Canada geese and sandhill cranes fill the skies during spring migration in May, heading to staging areas in the Saskatchewan River Delta to the north. Bald eagles are often observed around the lakes from spring to freeze-up.

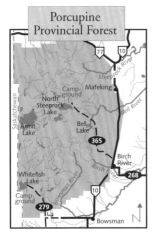

Access: The central part of the forest, and the campground at North Steeprock Lake, can be reached by following PR 365 northwest from PTH 10, which runs along the eastern boundary of the Provincial Forest. Whitefish Lake can be accessed by following PR 279 west of PTH 10.

Overflowing River

Perched on large boulders, American white pelicans spend a sunny afternoon at Overflowing River.

Flowing east out of the Pasquia Hills in west-central Saskatchewan, Overflowing River traces a sinuous pattern through thick forests to the northwestern tip of Lake Winnipegosis. As it nears its mouth, the river runs broad and shallow over a bed of rocks and rounded cobbles, twisting in a double S bend around a small wayside park.

These conditions – clear running water, boulders for perching and a wealth of small fish – draw birds by the hundreds. The park, with its shaded picnic sites and winding roadway, provides a perfect place to view the birds without startling them.

Pelicans, spotted sandpipers, herring and Bonaparte's gulls, as well as common and black terns perch on rocks or hover above the water, searching for fish. Visitors may see a great blue heron, nearly hidden by overhanging vegetation, as it waits motionless for minnows by the far shore. Downstream, cormorants by the dozen crowd the huge driftwood logs at the edge of the lake. During spring migration in May, several other species of sandpiper can be seen, and noisy flocks of sandhill cranes and Canada geese pass by overhead.

Mammals are also drawn to Overflowing River. Moose frequent the area, and can often be seen in late spring and early summer with spindly-legged young. Black bears frequent the riverbanks as well.

PTH 10 crosses Overflowing River about 100 kilometres south of The Pas. The wayside park is on the east side of the highway.

213

Tom Lamb & Saskeram WMAs

The mallard, above, is the most abundant duck in Manitoba. It arrives very early in spring, often when the lakes and marshes are still covered with ice.

These Wildlife Management Areas encompass much of the Saskatchewan River Delta near The Pas in west-central Manitoba. Tom Lamb WMA covers an expansive 214,459 hectares to the east and south of the town; Saskeram WMA ranges over 95,871 hectares to the west and north.

This is a low-lying area dominated by large shallow lakes and marshes, winding streams, potholes and muskeg. However, the delta also supports extensive stands of upland forest of balsam poplar, American elm, green ash and Manitoba maple, as well as willows.

Elevated limestone outcrops sustain white spruce, balsam fir and jack pine, mixed with aspen and birch. Because the soil is shallow atop the underlying bedrock, tree growth here is often stunted. It's not the forests that draw visitors to these WMAs, however. Rather, this is a place to explore extensive marsh habitats, most accessible only by boat.

Those most familiar with these areas (mainly hunters and wildlife biologists) say there is a "mosaic of wildlife" to be found here. The major large mammal species is moose, but white-tailed deer may also be seen along the levees. Woodland caribou occasionally visit the extreme northwest portion of Tom Lamb WMA.

Of the furbearers, muskrats are most numerous followed closely in number by mink and beavers. Other mammals include fishers, lynx, martens, raccoons, wolverines and red foxes of all colorations – red, blue, silver and cross. Black bears amble along the rivers and ridges. The variety of habitats also supports coyotes, several packs of wolves, skunks, weasels, red squirrels and jackrabbits.

These wetlands are major producers of waterfowl and stage vast numbers during migration. Ducks Unlimited has constructed several major water control works to optimize breeding and staging habitat in these areas. Ravens and bald eagles are common sights along the rivers and lakes, which are full of pike, walleye, suckers, whitefish, tullibee and goldeye.

Tom Lamb WMA: travel north from The Pas on PTH 10 until you reach PR 287. Turn east and travel past Clearwater Lake to PR 384. This road, more commonly called the Moose Lake Road, runs through the eastern part of the WMA to Moose Lake.

Saskeram WMA: access is by boat in summer and snowmobile in winter. Backcountry travel for experienced hikers with proper equipment is also possible through this challenging landscape. For further information on the area, contact the Department of Natural Resources office or the Ducks Unlimited Office in The Pas.

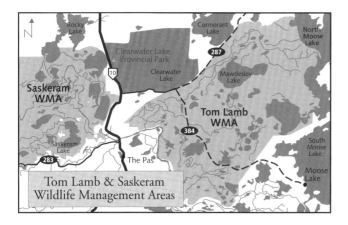

Despite its small size, the mink is an efficient marshland hunter, an accomplished swimmer and a fierce and fearless fighter. Like other members of the weasel family, it emits a pungent odor when provoked. Unlike its otter cousins, its toes are not webbed.

Photographer's Notes

June 21st: "This has to be moose heaven. And a great place for photographers too. With a little luck, I could have had a great blue heron, tundra swan and two species of ducks – all in one shot."

Northern Manitoba

Northern Manitoba
The Boreal Forest

Though Manitoba is better known for its southern plains and great lakes, the largest part of the province is boreal forest, a vast expanse of rock, water, trees and abundant wildlife. The ancient granite bedrock can be seen in cliffs and outcroppings throughout the region, but in many places glacial deposits – eskers, beach ridges and lacustrine silt – cover the rock. In the province's far north, eskers create natural causeways through tundra and taiga, while farther south the thick lake deposits nurture dense forests.

Broadly speaking, northern Manitoba can be horizontally divided into three bioregions. The more southerly boreal forest is dominated by thick stands of white spruce or jack pine interspersed with birch and poplar, with tamarack and black spruce in wetter areas. The transitional forest is in the middle, where stands of spruce and tamarack are thinnner and the ground is carpeted with moss and lichen. The transitional forest eventually gives way to treeless tundra in the far north.

The entry into the Precambrian Shield is perhaps most dramatic in Grass River Provincial Park, east of Flin Flon. Within a few kilometres of the park entrance, PTH 39 leaves the Manitoba Lowlands, with its characteristic limestone, and enters shield country, with its granite outcroppings, clear lakes and dense coniferous forests.

Jerry Kautz

From Grass River Provincial Park, left, to the shores of the Seal River, where a barren-ground caribou bull, opposite, pauses in his migration, Manitoba's northland is full of beauty, diversity and exceptional wildlife viewing opportunities.

Northern Manitoba

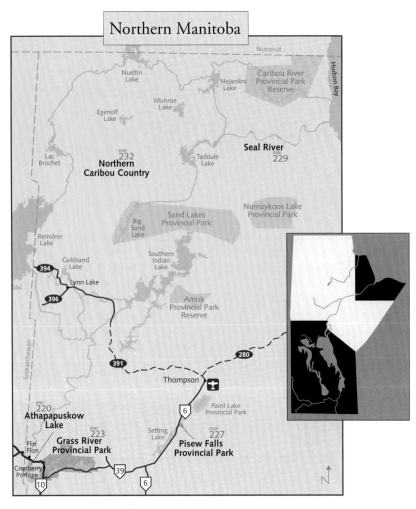

Chartered air service, right, is available to many northern lodges.

Farther east and north, the hard, resistant bedrock provides spectacles with water that can be viewed at Pisew Falls Provincial Park. One, for which the park is named, is located just off PTH 6, while another – Kwasitchewan Falls – requires hiking a 24-kilometre return trail along the Grass River.

In the province's far northern reaches, the highways give way to waterways and airways. For millennia, travel in both summer and winter was easier on lakes and rivers than on land. Even today, though challenging, canoeing Manitoba's northern rivers is still one of the best ways to see the province's unspoiled wilderness. The Seal, a Canadian Heritage River, combines stunning scenery and an ancient human history with remarkable wildlife viewing opportunities.

Even the most remote communities, lodges and parks can be reached within hours by air, making Manitoba's unparalleled northern wildlife easily accessible.

Trip Planning

If you have a day in the region . . . ❁ Spend it hiking at Grass River Provincial Park or Pisew Falls Provincial Park.

If you have two or three days . . . ❁ Take an excursion to Grass River Provincial Park or perhaps a weekend trip to Athapapuskow Lake. ❁ From Pisew Falls, hike to Kwasitchewan Falls and camp overnight by the rapids.

If you have a week or more . . . ❁ Spend it at one of many fly-in lodges. ❁ Canoe the Grass River, one of many paddling options. ❁ Like to drive? The 2,000-kilometre Northern Woods & Water loop takes you from Winnipeg to Thompson and includes many of the sites featured in this and other sections. Head north on PTH 6 to Eriksdale, then PTH 68 past Dog Lake and over the Lake Manitoba Narrows (Central section) to Dauphin and Vermillion Park (Western section). Continue up PTH 10 to Duck Mountain and Porcupine Forest and spend some time at Grass River Provincial Park and Athapapuskow Lake. Head east on PTH 39 to rejoin PTH 6 up to Thompson. Stop at Pisew Falls and Paint Lake, returning to Winnipeg on PTH 6 with stops at Long Point and Mantagao Lake (Central section).

Northern Manitoba

Athapapuskow Lake

Athapapuskow is a large, deep, oxygen-rich lake near Flin Flon in northwestern Manitoba. Covering some 270 square kilometres, a small portion of the lake spills across the Saskatchewan border. Its clear, cold water provides an ideal habitat for lake trout.

Athapapuskow, along with Nueltin and Clearwater Lakes, tend to provide the largest lake trout for anglers. This is largely due to careful management and the implementation of mandatory release of all trout over 55 centimetres.

The topography of Athapapuskow is similar to nearby Grass River Provincial Park, with features from both the Precambrian Shield and the Manitoba Lowlands.

Guy Fontaine

Mink are seldom found far from water and are excellent swimmers. They prey on fish, snakes, rabbits and even muskrats.

Twenty-kilogram lake trout cruise the waters of the south basin and 15-kilogram trout are a common occurrence. In the fall, especially early to mid-September, hundreds may be seen spawning in the rocky shallows and reefs, particularly along the southern shores of the lake. To protect the fish, a large part of the southern half of the lake is closed to angling between September 15th and the Canadian Thanksgiving weekend in mid-October.

The boreal forest and Manitoba Lowlands habitat surrounding the lake supports many large mammals. Moose are fairly common along the shoreline and woodland caribou are seen on occasion.

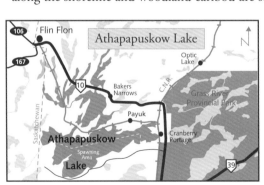

Outcroppings of dolomitic limestone are found along roadcuts; look closely and you may find the fossils of creatures that once found a home in ancient Manitoba's shallow inland seas.

Several wolf packs roam the area. Mink, beavers and otters are common in the lake's many bays and reaches.

Colonial nesting birds such as double-crested cormorants, great blue herons and American white pelicans can be seen on the lake between May and August, when they begin to move south. Waterfowl use the lake as well, with common goldeneye, lesser scaup and wigeon most frequently seen. Common loons nest on the lakeshore and the central basin attracts large numbers in fall prior to migration. Ospreys and bald eagles are also common from spring through fall.

Access Athapapuskow Lake can be accessed from several points along PTH 10. One of the best places is the provincial campground just north of Cranberry Portage. Bakers Narrows, Flin Flon and Cranberry Portage offer a selection of accommodations and services for wildlife enthusiasts and anglers.

Wolves, such as this black beauty, above, are synonymous with wilderness.

Northern Manitoba

Grass River Provincial Park

Located east of Flin Flon, Grass River Provincial Park has two geological regions that visibly split the park in half. Clean lakes, one of the best canoe routes in North America, and the possibility of seeing woodland caribou make a trip to this beautiful northern Manitoba park particularly worthwhile.

Just a few kilometres inside the park's southwestern boundary, the dolomite of the Manitoba Lowlands gives way to the granite terrain of the Precambrian Shield. In low-lying areas, small lakes and wetlands can be seen, while rocky outcrops in the larger lakes are typical of the shield. Glacial debris covers both kinds of rock.

Fens – marshes where the water is near the surface and slow moving – are filled with willows, alders, sedges and mosses. Peat deposits throughout the park act as natural filters, resulting in lakes and rivers that are exceptionally low in sediments.

Grass River Park is well known for its fish. Pike, walleye and whitefish abound in all the lakes, while lake trout are found in Reed and Second Cranberry Lakes. Black bears and river otters can be seen feeding on white suckers near the Karst Spring south of Iskwasum Lake. Moose are also common in the park. In May and June cows and young are often seen along the road and later in the season you may see a heavy-antlered bull feeding along the river.

Common loons nest on many of the lakes, as do herring gulls and common terns, which can be seen on small rocky islands on Reed, Leak and Second Cranberry Lakes. Double-crested cormorants and American white pelicans are also seen here.

Three-toed woodpeckers forage for insects among the towering spruce in the forests and boreal chickadees, gray and blue jays,

Jerry Kautz

The Grass River runs through rocky channels, tumbles over rapids and widens into lovely stretches of quiet water or lakes, providing superb canoeing and wildlife viewing opportunities.

In early June, calypso orchids, or Venus slippers as they are often known, can be found in open shaded areas of coniferous woods.

Bald eagles, or their enormous nests, can often be seen along the shorelines of the park's lakes or rivers.

golden-crowned and ruby-crowned kinglets, red-breasted nuthatches, white-winged crossbills, red crossbills, pine siskins and evening grosbeaks can all be seen. Boaters will often see the huge nests of bald eagles in the trees along the shorelines, with the adults nearby. Common mergansers, too, are a frequent sight.

In May and June, sharp-eyed visitors may catch a glimpse of woodland caribou on the park's larger islands, which are generally free of predators. These islands provide important calving sites. Great blue herons grace the lakeshores and several wolf packs roam the park like silver shadows.

Access By road, The Pas is 600 kilometres northwest of Winnipeg and 550 kilometres north of Brandon. From The Pas, the park is about 75 kilometres north via PTH 10 and 39. The latter runs through the heart of Grass River Provincial Park and past the largest lakes. Canoeists can put in at any one of three canoe launching sites on Simonhouse, Iskwasum or Reed Lakes. You can spend several days on the water, ending at the pictograph site on the northwest shore of Tramping Lake, just east of the park.

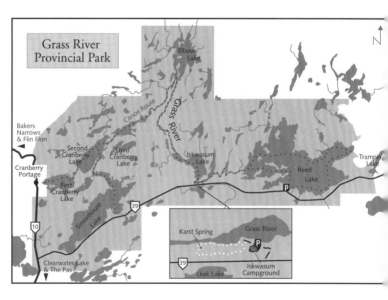

Karst Spring Trail

Karst formations, underground caves and channels created by water dissolving limestone, underlie much of the Manitoba Lowlands. A particularly picturesque example can be found near Iskwasum Lake, where a spring flows from a cliff to join the Grass River. The underground channel is believed to drain Leak Lake and flows north beneath PTH 39 to its outlet above the river.

Karst Spring pours from a limestone outlet, turning the surrounding area into an emerald haven.

Photographer's Notes

June 23rd: "A gem of a trail. Follows the river just far enough into the woods to allow a view of wildlife on the river."

Following the Grass River upstream, the 3.2-kilometre Karst Spring Trail takes you through boreal forest along a path carpeted with mosses and lichens. In boggy areas, the breeze carries the fleeting scent of Labrador tea. In early June, pink calypso orchids bloom along the trail, which climbs a ridge between the Manitoba Lowlands and Precambrian Shield. Halfway up the trail, Karst Spring spouts from a cliff to join the river below. Flowing year round, the water literally seems to come from nowhere. However, listening carefully at several places along the 3.2-kilometre trail, you can hear the current passing through the rock under your feet.

 The Karst Spring Trail starts at Iskwasum Campground.

Northern Manitoba

Pisew Falls Provincial Park

It is the sound that first pulls you to Pisew Falls – a distant thundering both heard and felt. As you near, the din grows louder. When at last you arrive at the falls the sound is so all-encompassing that it's almost forgotten, as a torrent of water plummets 13 metres to the river below.

Part of the Grass River system, Pisew Falls is the highest easily-accessible natural waterfall in Manitoba – the dramatic result of a fault in the Thompson Nickel Belt. For years, a visit to the falls meant a welcome break in the long journey from Winnipeg to Thompson.

A newly erected suspension bridge, built with funds raised by the Thompson Rotary Club and private donors, spans the Grass River, allowing a bird's eye view of the rapids just below the falls.

Beyond the bridge, a new backcountry hiking trail leads to Kwasitchewan Falls, a waterfall higher than Pisew. The trail, which is 24 kilometres return, is designed for fit and experienced trekkers and includes backcountry campsites.

Pisew Falls is surrounded by typical boreal forest of jack pine, white and black spruce, tamarack, birch and aspen, as well as speckled alder. In the immediate proximity of the falls are plants usually more at home in a wetlands setting – sedges and grasses, marsh marigolds, wild mint and palmate-leaved coltsfoot.

Although the falls are beautiful year-round, in winter the area is transformed into an ice palace, glittering from top to bottom with hoar frost and ice crystals.

The walls of the palace consist of great stands of white spruce, which have attained an impressive size after decades of high humidity. They drip with a coat of lichen that in winter dons an extra coat of ice and snow, turning them into natural ice sculptures. About their feet are what appear to be giant shaggy bushes. In fact, they are ice-coated branches that break and fall about the trees due to the buildup of heavy ice. Ice bridges form

At Pisew Falls, opposite, the Grass River thunders over a fault in the Thompson Nickel Belt.

Pisew is the name early Manitobans gave to lynx. Though rarely seen, the lynx is completely at home in the northern forests. It swims well, climbs well and moves with ease through dense forests. It is also perfectly adapted to winter, with broad, well-furred feet that allow it to move easily through deep snow.

Seen here from the air, Kwasitchewan Falls is higher than Pisew.

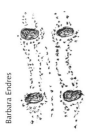

Sleek and playful, river otters are also wary and only seldom seen, but their tracks, above, or the pattern of the tracks are often evident.

over the Grass River, seemingly created out of thin air. In fact, the ice is a combination of spray and mist hardened to a solid arc by extreme winter temperatures.

During other seasons, fungus lovers will enjoy ferreting through the dark, damp forest floor identifying a myriad of mushrooms, toadstools and other fungi – as well as numerous mosses – all mixed with an abundance of rock-cranberry and bunchberry. The ancient rock outcrops are covered with reindeer lichen, common throughout the area.

A quiet stroll along the trail may be rewarded with a glimpse of a black bear, a moose or mink, perhaps even a marten. Almost certainly there will be signs of small mammals like red squirrels and snowshoe hares. In winter, otters can sometimes be seen sliding on the snowbanks along the river.

In spring and summer, gray jays, boreal chickadees, golden-crowned and ruby-crowned kinglets, pileated, black-backed and three-toed woodpeckers, as well as northern hawk-owls and boreal owls can be seen. Both ruffed and spruce grouse are also found.

 Pisew Falls is located 65 kilometres south of Thompson on PTH 6.

Seal River

Of Manitoba's major rivers, the Seal is one of the largest undeveloped systems, flowing freely across Manitoba's far north and emptying into Hudson Bay. Few people venture down this Canadian Heritage River each year, but those who do will find pristine wilderness and dozens of species of birds and mammals.

The Seal has been used by people for thousands of years as a travel route, a place to hunt migrating caribou, and a source of plentiful fish. Campsites, some more than 6,000 years old, have been found on eskers in the region. These ridges of sand and gravel, deposited at the end of the last glaciation by meltwater rivers tunnelling through the glaciers, often wind across the landscape for many kilometres. As much as 30 metres high and 300 metres wide, they provide relief in a terrain devoid of hills and have served as natural thoroughfares for millennia.

Some eskers are dotted with spruce, others are as naked as the rocks and make great hiking trails. But, in September, hikers may have to yield these wilderness highways to the herds of barren-ground caribou that use them as migration routes.

The North and South Seal Rivers join north of Tadoule Lake in the boreal forest. From here, the Seal River flows north through transition forest. Marked by stunted spruce and tamarack with flattened club-shaped tips, this habitat is known as the "Land of Little Sticks".

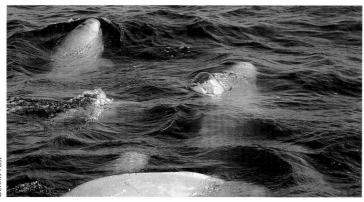

Beluga whales congregate by the thousands in the Seal River estuary every summer.

Like others of this family, the rare northern or Franklin's lady's slipper has no nectar in its slipper-like pouch, but attracts pollinating insects with a nectar-like scent.

The polar bears of Hudson Bay spend seven or eight months a year hunting seals on the ice. In summer, however, they are forced ashore by melting ice where they feed little, if at all, surviving on fat reserves built up over the winter.

The river then continues east through the transition forest to Hudson Bay. During its journey to the sea, the Seal tumbles over challenging rapids and fields of boulders called felsenmeer that give some stretches of the river a moonscape quality. It also rushes through scenic gorges, and over sandspits and bars. Along the shores, wildflowers and mosses skirt the banks and cover the adjoining lands in a bright palette of color during the brief, bright summers (June through August).

More than 160 kilometres inland from Hudson Bay, close to the 59th parallel, harbor seals play and feed in Shethanei Lake. The river may be the best place in the world to see this freshwater adaptation by a marine mammal, for the seals can be seen in any month of the year. During the winter months, they congregate around the rapids, where the force of the water keeps the ice at bay. Polar bears can also be seen many kilometres inland during the summer, as well as along the Hudson Bay coast from June to October, waiting for the bay to freeze. More than 3,000 beluga whales spend the summer in the river's estuary. They are best seen during June and July.

Wolverines hunt along the banks and golden and bald eagles soar overhead, searching for voles, hares and the Arctic ground squirrel or siksik, found here at the southern extent of its range. In winter, many of the 400,000 strong Qamanirjuaq barren-ground caribou herd roam well-worn trails on the spines of the eskers of the Seal River valley, trailed by wolves.

In the channel south of Great Island stands Bastion Rock, a dramatic 20-metre-high monolith. A colony of cliff swallows

Tidal pools reflect a brilliant June sky along the estuary of the Seal.

nests on the rock, the northernmost limit of the species' range. Buff-breasted sandpipers and American black ducks are common in summer. Red-throated loons are more common here than in Churchill. Ross's geese are abundant in August and short-billed dowitchers, American golden plovers and semipalmated plovers can be seen here in migration.

If mosquitoes and black flies can be considered wildlife, they can be found here in abundance in July and August. Prepared travellers will wear bug hats, shirts, jackets and/or repellent.

Access There is no road access to the Seal River. The community of Tadoule Lake, on the lake of the same name, is the only permanent settlement along the river. To travel the Canadian Heritage River portion of the route by canoe, fly to Tadoule Lake direct from Winnipeg, or from Thompson or Gillam. Put in on the South Seal, follow that tributary 35 kilometres northeast to the Seal River, then travel downstream for 260 kilometres to Hudson Bay. Arrangements for coastal air or boat charter from the mouth of the Seal to the town of Churchill must be made before departure. Because of the challenging nature of this waterway, this route should be approached with caution and thorough advance preparation. Several fly-in lodges in northern Manitoba can arrange canoe trips on the Seal.

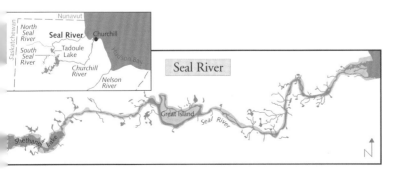

Northern Manitoba

Northern Caribou Country

Manitoba's vast northland is almost as much waterscape as landscape. This is where the province makes good its boast about being home to 100,000 lakes, fed by innumerable creeks, streams and great rushing rivers. Across this vast, sparsely-populated territory, the terrain is challenging and ever-changing. This diversity is celebrated in two provincial parks and two provincial park reserves totalling 2.1 million hectares.

The boreal forest of Amisk Provincial Park Reserve, south of the 57th parallel, gives way to long sand beaches and transitional forest in Sand Lakes Provincial Park, which straddles the 58th parallel. To the east, Numaykoos Lake Provincial Park includes both the Precambrian Boreal Forest and Hudson Bay Lowlands natural regions, while Caribou River Provincial Park Reserve comprises transitional forest and Arctic tundra. The last of these lies along the province's northern border with Nunavut.

The layer of skin, or velvet, covering this bull caribou's magnificent antlers contains blood vessels that carry nutrients necessary for growth.

Amisk, the Cree name for "beaver", has not only beavers but a large number of moose as well. Big Sand Lake in Sand Lakes Park is noted for its breeding population of Caspian terns. Caribou River features the ancient shore-line of the glacial Tyrrell Sea and a huge crater lake – Round Sand Lake – created by a meteorite. Numaykoos ("trout" in Cree) has plentiful fish in the watershed of the Beaver River, and a variety of terrestrial species from pine martens to ptarmigans.

Manitoba's north country includes far more than these protected areas. Beyond the park boundaries, hundreds of deep cold lakes harbor huge northern pike and lake trout, bald and golden eagles soar overhead, and wolves prowl the shores. In late spring, particularly in the province's northern reaches, the land comes alive with breeding Canada geese, tundra swans, sandhill cranes, common loons and Arctic terns. Nueltin Lake, on the province's northern border, supports a large population of white-winged scoters, while black scoters are seen during migration.

Dennis Fast

Eskers, like this one near Nueltin Lake, snake through northern Manitoba, creating a natural highway system that has been used for millennia.

Most of all this is caribou country. For millennia, the Inuit and Dene have gathered in late August at well-trodden river crossings near the Nunavut border, pinning their hopes and their lives on intercepting the vast migration of barren-ground caribou. When the animals came, it was as a river of life that meant food and clothing to survive the long winter. Even today, the herds sometimes take days to pass a given point, a spectacle rivalling those of the African plains.

The autumn migration often follows eskers that wind through northern Manitoba. Remnants of the last glaciation, these high ridges of sand and gravel serve as natural highways through the tundra, leading the animals deep into the transition forest. Here the earth is carpeted with reindeer lichen, providing winter food for the caribou. The Qamanirjuaq herd, nearly a half-million strong, ventures as far south as Sand Lakes Provincial Park.

In spring, the migration is reversed. In March, the animals gather on ice-covered lakes, sometimes in groups of 10,000, to begin their long journey north.

Access There are no year-round roads into Manitoba's far north, but air connections reach a number of small communities and several fly-in lodges offer opportunities to experience wildlife in each of these natural regions. Call Travel Manitoba or the Manitoba Lodges and Outfitters Association for information.

Fast Facts

Though the plains grizzly disappeared from Manitoba in the 1880s, a cousin, the barren-ground grizzly, occasionally makes an appearance in the province's far north. Its range coincides with that of the Arctic ground squirrel, or siksik, which is seldom seen south of the Seal River.

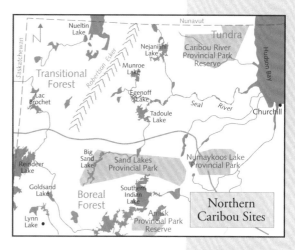

Northern Caribou Sites

Coastal Manitoba

Coastal Manitoba
The Hudson Bay Lowlands

Manitoba's Hudson Bay Lowlands form the western edge of an enormous saucer-shaped basin surrounding Hudson Bay, a shallow inland sea that penetrates deep into northern Canada. The basin was flooded about 8,000 years ago following the retreat of the continental ice sheet after the last glaciation.

Since that time, the land around the bay – relieved of its great weight of ice – has been rising at a rate of about 80 centimetres every century. This isostatic uplift is dramatically demonstrated at Sloop Cove National Historic Site on the west side of the Churchill River estuary. Here, in the 1760s, the Hudson's Bay Company secured small sailing boats to metal rings in the shoreline rock. Today, the cove is a meadow.

Much of the land now risen from the sea is a lowland plain, locked in permafrost. Dotted with ponds and marshes, its vegetation includes Arctic plants not seen elsewhere in the province, as well as stunted spruce, aspen, willows and birch inland from the bay. This is caribou country and the region supports two coastal herds of caribou. A number of large rivers – including the Seal, Churchill, Nelson and Hayes – empty into the bay's western side. The tidal estuaries at the mouths of these rivers attract many species of marine mammals, including whales and seals.

From December through June, the region's well-adapted population of polar bears hunts seals on the pack ice. When melting ice

Though its luxurious white coat makes it seem larger, the Arctic fox is one of the smallest members of the canine family in Canada (only the swift fox is smaller), about the size of a house cat. Its vocalizations set it apart as well. Instead of a bark, the call of the Arctic fox is a high-pitched undulating whine.

Bonaparte's and ring-billed gulls wheel above the surf at the mouth of the Churchill River.

Coastal Manitoba

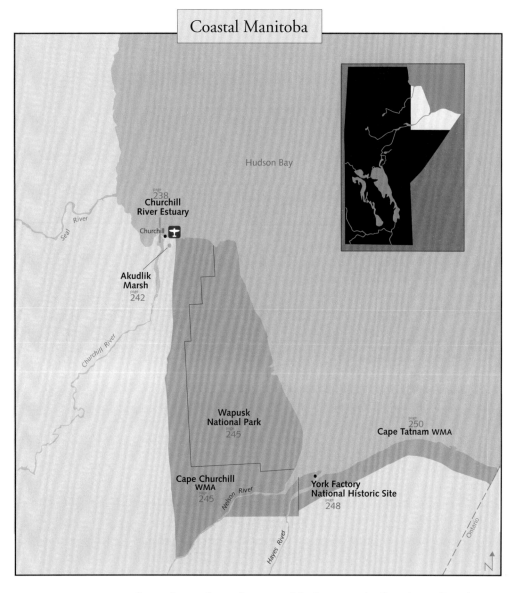

The beauty of the Churchill region draws photographers, right, from around the world.

forces them ashore, the great white bears can be found prowling the shores of Cape Churchill WMA, Wapusk National Park and Cape Tatnam until mid- to late November.

Though challenging, this is an environment rich in wildlife and largely unspoiled by development. For wildlife watchers willing to meet its challenges, the rewards can be remarkable.

Trip Planning

If you have a day in the region . . . ◎ Spend it at Cape Churchill WMA or on the Churchill River Estuary on a one-day summer beluga whale or autumn polar bear safari with licensed tour operators (one-day return air excursions can be arranged from Winnipeg).

If you have two or three days . . . ◎ Explore Akudlik Marsh (May to September), the Churchill River Estuary (June to August) and Cape Churchill WMA (May to August for birds, late September to late October for polar bears).

If you have a week or more . . . ◎ Book a birding (May/June), beluga whale (July/August) or polar bear (October) excursion to Churchill and visit Cape Churchill WMA, the Churchill River Estuary and Akudlik Marsh. ◎ Fly from Winnipeg to Cape Tatnam and spend a week at a lodge on the coast of Hudson Bay, viewing polar bears, caribou and many species of birds. ◎ Fly to Gillam and take a tour down the Nelson River to York Factory National Historic Site and the Hudson Bay coast, where whales, seals, polar bears and caribou can be seen. ◎ Consider a week-long winter excursion in Cape Churchill WMA that allows you to experience the life of a trapper.

Coastal Manitoba

Churchill River Estuary

The basalt rock of Cape Merry bears glacial striations, a legacy of the last ice age. The appearance of the rock is softened in summer by masses of wildflowers.

The town of Churchill sits on Hudson Bay's western shore near the mouth of the Churchill River. Across the estuary the star-shaped Prince of Wales Fort sits on a finger of land, its cannons still pointed seaward.

Churchill is famous for its polar bears. In October crowds of visitors come to Cape Churchill WMA east of the town to see the great white bears as they congregate along the coast, waiting for the bay to freeze.

But there is more than bears to see in Churchill. Here, in spring and summer there are superb opportunities to see Arctic birds, whales and wildflowers, as well as national historic sites on both sides of the river. In winter, Churchill's magnificent northern lights draw people from around the world.

Cape Merry National Historic Site, the rocky promontory at the mouth of the Churchill River, is a favourite of birders as well as history buffs. In late June, as the bay ice breaks up, the water beyond the cape is filled with large rafts of sea ducks, eiders, oldsquaw and scoters, along with smaller flocks of Pacific loons. In July, beluga whales and ringed and harbor seals come and go with the tides. Gulls, terns and jaegers soar overhead.

Farther inland shallow freshwater ponds are filled with a great variety of birds in migration. Here, Arctic terns and Bonaparte's gulls are abundant. In early June, it's possible to see Thayer's, Sabine's and little gulls. On the river, white-winged, surf and black scoters can be seen, along with oldsquaw ducks.

At the end of the road at Cape Merry are the remnants of an artillery battery erected in 1746 to provide added protection for Prince of Wales Fort across the river. Boat tours cross the two-kilometre river mouth to the fort in July and August.

In July, this is one of the best places in the world to view beluga whales, the most vocal of whales. After the ice breaks up, beluga whales return by the thousands to feed in river estuaries along the coast of Hudson Bay. It is estimated that some 3,000 belugas return to the Churchill River. Pods of these snowy white adults and gray-colored young swim around the tour boats, some of which are equipped with hydrophones to eavesdrop on the belugas' underwater songs.

Prince of Wales Fort is a huge, partially-restored 18th-century stone fort. Built by the Hudson's Bay Company to protect British interests in the fur trade, construction took more than 40 years. Three kilometres upstream, accessible by boat or trail from Prince of Wales Fort, is Sloop Cove National Historic Site, an 18th-century harbor for Hudson's Bay Company sailing boats, complete with iron mooring rings. Because of post-glacial land uplift, the area is now a meadow embraced by rocks.

The area around the fort is also great for birding. In mid-June, there are king eiders and long-tailed jaegers. Ross's gulls have been sighted here, but they are more often observed at Akudlik Marsh.

Dennis Fast

Hudson's Bay Company officer Samuel Hearne carved his name into the rock at Sloop Cove.

Complete with cannons, Prince of Wales Fort was constructed in the 18th century to defend Hudson's Bay Company fur trade interests. However, it was surrendered to the French without a shot being fired in 1782.

Coastal Manitoba

The long, warm days of June and early July carve magnificent ice sculptures along the shores of Hudson Bay.

Churchill is accessible by air or rail from Winnipeg or Thompson (the 1,600-kilometre railway line was originally built over permafrost and muskeg in 1929 by a crew of 3,000). Tour boats offer whale excursions during the summer months. Several companies offer excursions across the river to Prince of Wales Fort.

Polar bears, like the rock climber opposite, can be more than three metres tall when standing on their hind legs, weigh in excess of 450 kilograms and run faster than 50 kilometres an hour. When hunting, they wait patiently near breathing holes for an unsuspecting seal to appear. The Churchill area population is about 1,200.

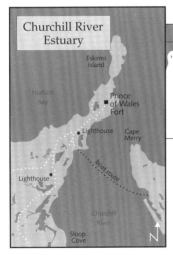

Akudlik Marsh

Bearing one white, five-petalled blossom atop each stalk, and a single leaf partway up the stem, the northern grass-of-Parnassus flowers in late summer.

Bounded on the north and east by the basalt bedrock of Hudson Bay, Akudlik Marsh is a 559-hectare maze of ponds and shallow lakes, used by many birds as a nesting and breeding site.

In Inuktitut, Akudlik means "the place in between" and refers to the deserted village site south of the marsh. The small settlement was built between Churchill and the airport during the 1950s to house workers for the Northwest Territorial Government. When the government headquarters moved north to Rankin Inlet, the workers left.

In 1980, when a pair of rare Ross's gulls nested here, Akudlik Marsh became the first known North American nesting site. Since then, the marsh has become a very popular birding location, so much so that a Special Conservation Area has been established to protect the nesting gulls. Public access is restricted once nesting sites are identified, but adult birds are often seen feeding in the vicinity.

Just north of the old Akudlik settlement, an interpretive trail has been established on dikes built to manage water levels in the marsh. From the end of May to early July, the wetlands are also frequented by Pacific loons, horned grebes and Hudsonian godwits. Semipalmated plovers often nest along the dikes.

On the north side of the marsh, Thayer's and herring gulls can often be seen from the Coast Road, which runs along a rocky ridge

A bird that likes company, several adult male common eiders sunbathe on an ice floe. In winter, many do not migrate, but search out open water patches on the bay and remain year-round.

The neck band or collar of the semi-palmated plover darkens during the breeding season, then fades from black to brown during the winter.

beside the bay. Farther east, a narrow (and often soft) road runs south to Lions Club Park. Here, in a grove of gnarled spruce, merlins can sometimes be seen.

The shallow marsh lakes are great places to see shorebirds up close, including white-rumped or Baird's sandpipers, sanderlings, lesser yellowlegs, short-billed dowitchers and red-necked phalaropes.

Just northwest of the marsh, short-eared owls can often be seen from the road, sitting atop rocks, and American golden-plovers nest in the wet meadows below the ridge.

Access Akudlik Marsh can be reached by following Churchill's main street (officially Kelsey Boulevard, but generally called The Highway) south-east. The interpretive trail is four kilometres from town. Alternatively, the Coast Road (which can be blocked by snow as late as the end of June) circles the north and east side of the marsh. Walkers should remember that polar bears can be found here even in the summer months and escorted tours are recommended.

Guy Fontaine

The rare Ross's gull wears a black neck band during breeding season. In flight, both sexes can be distinguished by their rosy breasts.

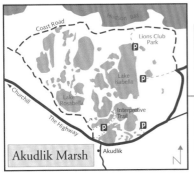

243

Coastal Manitoba

Wapusk National Park
& Cape Churchill WMA

One of Canada's newest national parks, Wapusk, meaning "white bear" in Swampy Cree, was established in 1996. Twice the size of Prince Edward Island, it stretches along the western shore of Hudson Bay, next to the Cape Churchill Wildlife Management Area.

Extending far inland and representing the lowlands of Hudson Bay and James Bay, the national park comprises freshwater ponds, marshes, fens, bogs and transitional forest. The combination attracts bountiful wildlife. Waspusk protects one of the world's largest known polar bear denning sites. It also provides habitat for a large herd of caribou, as well as foxes, wolves, wolverines and fishers. The park also provides critical habitat for waterfowl and shorebirds during the nesting season and seasonal migrations.

Cape Churchill WMA, which borders the park's western and southern boundaries, aids in protecting the fragile coastal ecosystem and the wildlife it supports, while providing an area accessible to wildlife watchers. The WMA, which is Manitoba's largest, has become world-famous among naturalists and photographers for research and wildlife viewing.

Though caught here in transition from winter to summer attire, the Arctic hare, left, is almost always dressed for the season.

Wapusk is a huge lowland plain, where peat bogs and wetlands occupy most of the region and conifers are confined to ridges.

Coastal Manitoba

The Bonaparte's gull is a bouyant flier and can be mistaken in flight for a tern. It is the only gull species that regularly nests in trees.

Geologically, a limestone plain underlies both the national park and the WMA, covered by the most extensive mantle of peat in North America. This was the virtual centre of the last glaciation and the enormous weight of the continental ice sheet depressed the Earth's crust. Since the ice melted about 8,000 years ago, the crust has slowly sprung back like a depressed sponge. Every century, it rises about 80 centimetres.

People have lived along the coast of Hudson Bay for thousands of years and special provisions are made in Wapusk National Park to allow the continuation of many traditional land uses, including fishing, trapping, hunting and the gathering of wood and berries.

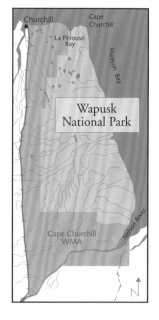

Unlike most Arctic and subarctic ecosystems, the Cape Churchill area is on the edge of the boreal forest and boasts more than 400 plant species. The climate is severe, but plants seem to know they don't have a lot of time to flower and reproduce. From mid-June through July the ground is a palette of changing colors as different plants bloom. White mountain avens are the first, followed by a succession of delicate wildflowers. Bearberry changes the landscape to red with the first touch of frost.

The area may be best known for polar bears, but there is other wildlife here, including caribou, which may sometimes be seen on the tidal flats or wading out into the bay in summer, escaping the mosquitoes and flies. In May or June, ringed seals and harbor seals, as well as smaller numbers of bearded seals can be seen.

In July and August, thousands of beluga whales (see Churchill River Estuary in this section) return to the Nelson, Churchill and Seal River estuaries.

Birders travel thousands of kilometres to see the rare Ross's gull and the graceful Arctic tern, as well as roughly 200 other species. Tundra swans, Arctic loons, peregrine falcons and gyrfalcons can also be seen along with fox sparrows, red-necked phalaropes, stilt sandpipers, pine grosbeaks, parasitic jaegers and willow and rock ptarmigan. Snow geese and Canada geese are a common sight all along the Hudson Bay coast. The tidal flats and salt marshes of La Pérouse Bay, just west of Cape Churchill, is Manitoba's most famous breeding ground for snow geese.

Butterflies, such as Polaris fritillary and Polixenes arctic, flit over the tundra near Churchill, the only place in Canada other than Yukon where one can see these tiny beauties.

Access Churchill is served by regularly scheduled commercial airline or railway. There are few roads in the Cape Churchill WMA and none in the national park, but tour operators provide special tundra vehicle excursions in the WMA or aerial tours of the WMA and national park. Wapusk National Park and Cape Churchill WMA are rugged, unserviced wilderness areas. Contact Parks Canada personnel in Churchill about the national park or Travel Manitoba for information on tour operators.

Dennis Fast

In summer, polar bears leave their enormous prints in the sand along Hudson Bay. In October and November, they gather near Churchill in one of the world's densest autumn concentrations, before moving out on the ice in search of seals. Pregnant females move inland to denning areas in Wapusk National Park, Cape Churchill WMA and Cape Tatnam WMA. They give birth to their tiny cubs between late November and mid-January.

A tranquil pond belies the rush of life that is summer in Manitoba's north.

Coastal Manitoba

York Factory,
the Nelson & Hayes Rivers

Purple paintbrush can be found in well-drained open areas.

For centuries a bustling fur trade centre, York Factory National Historic Site now sits in lonely splendor, its surrounding marshes and forests once again the domain of the region's wildlife.

York Factory National Historic Site rests on a peninsula between the mouths of the Nelson and Hayes Rivers on the western coast of Hudson Bay. Situated five miles upstream from the bay on the north shore of the tidal estuary of the Hayes, it is a solitary reminder of more than 250 years of fur trade history.

Here, summers once rang with activity as the Hudson's Bay Company fur trade brigades, burdened by beaver pelts in the thousands, came downstream from the far reaches of Western Canada. Just offshore, avoiding the treacherous shoals of the river's mouth, the company ship from London anchored each August, dispensed a year's worth of supplies and loaded the pelts on board for the journey back to England.

Wildlife provided sustenance for the employees of York Factory in those days, but today, the land again belongs to the birds, the bears, the caribou and the seals. Salt marshes, created by the ebb and flow of the tides, stretch east of the old fort, and stunted spruce can be found to the north, between the two rivers.

Though the Hayes and Nelson trace nearly parallel courses from the north end of Lake Winnipeg north and east to the bay, the rivers have very different characters. The Nelson, now dammed, was once wild and all but unnavigable. The Hayes is a series of connected waterways. Though it was longer and presented fur

traders with more than 30 portages, it was the route of choice to the Canadian interior. Both rivers tumble off the Precambrian Shield upstream far west of the bay, and have large tidal estuaries that attract bountiful wildlife.

Dennis Fast

Readying itself for winter, a willow ptarmigan molts its summer feathers in favor of snowy white plumage.

It is the combination of habitats that makes wildlife viewing so extraordinary here. Black bears roam the area during spring, summer and fall, and furbearers such as foxes, martens, wolves, fishers and wolverines hunt year-round. The occasional polar bear may also be seen in August, making its way along the coast.

Caribou often spend their summers on the many islands in the Hayes. In the river estuaries, bearded seals — huge 360-kilogram giants – as well as smaller ringed seals can be seen, along with beluga whales.

During the summer months, small coastal herds of caribou feed on the lichen, sedges and woody plants north and east of the old fort, while moose are year-round residents here. During the summer months, visitors may see golden and bald eagles, ospreys and short-eared owls. This area also draws great numbers of both Canada and snow geese during spring and fall migrations.

In winter, moose can often be found on the islands, which are blanketed by aspen and birch. Snowy owls and willow ptarmigan can be seen year-round and winter brings rock ptarmigan, snowy white except for a black tail marking and the black eye line of the male.

Access There are no roads into York Factory, but it is accessible by air and by water in the summer. Air charter transportation can be arranged from several places in northern Manitoba. A lodge at York Factory can arrange boat transportation from a landing strip on Hay Island to York Factory National Historic Site. Contact Parks Canada in Churchill for more information.

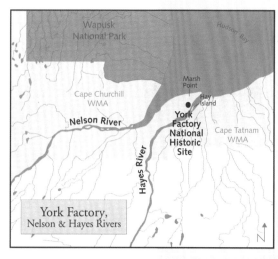

York Factory, Nelson & Hayes Rivers

Cape Tatnam WMA

Almost perfectly aligned in flight, a rare formation for this species, snow geese head for Cape Tatnam. There they stage in large numbers in September, before continuing their journey south.

Like a ribbon of rare beauty, Cape Tatnam Wildlife Management Area traces the shoreline of Hudson Bay from the mouth of the Nelson River east to the Manitoba-Ontario border. Comprising just over 530,000 hectares, this is Manitoba's second-largest WMA. Only Cape Churchill WMA to the north is larger.

This is northern wilderness that fuels the imaginations of would-be adventurers. Today, you'll find three fly-in lodges, one of which caters to ecotourists as well as goose hunters.

The coastal habitat lies along the clear waters of Hudson Bay, with seemingly endless stretches of sand beaches relieved here and there by high ridges and broad rivers, such as the Kaskattama River.

Away from the coast – heading southwest – is a broad, treeless 10-kilometre band between Hudson Bay and the tree line. Here, visitors can see a series of raised beaches, a reminder of the past when Hudson Bay stretched inland at least 150 kilometres. Like Wapusk National Park to the northwest, this region is a polar bear denning site, where females move inland in autumn and give birth to their cubs in winter.

The vegetation here includes birch, willows and tamarack, with Labrador tea, sedges, grasses and lichen underfoot. Farther inland from the bay, the vegetation gives way to a subarctic transitional forest. Here, white and black spruce, tamarack and jack pine grow thickly, though smaller in size compared to their southern cousins.

Cape Tatnam has a growing international reputation for its wildlife, including polar bears, caribou and snow geese. The best viewing period for polar bears is in July, August and September. During these months the bears gather along the shores to pass the summer, then migrate north to Churchill to await freeze-up.

Summer is the best time to see caribou. In June and July the herds roam the tidal flats to escape the black flies and mosquitoes. In fall, they move inland, grazing on lichen, sedges and woody plants as they fatten up to winter in the woods.

This is a birders' paradise. Willow ptarmigan, snowy owls, bald eagles and ospreys are all found here. Parasitic jaegers are commonly seen and long-tailed jaegers, the smallest of the species, are here at the southern edge of their range. Bonaparte's gulls, with their flashy white wing tips can be seen here, along with Arctic terns. The northern shrike may also be found.

Hundreds of thousands of snow and Canada geese, and smaller numbers of Ross's geese arrive in September to rest and feed before travelling farther south.

Black bears can be found at Cape Tatnam, along with moose and many fur-bearers, such as martens, wolverines and wolves.

Seaward, there are opportunities to see beluga whales, 360-kilogram bearded seals and ringed seals in great numbers.

Air access to Cape Tatnam can be arranged through one of the lodges. Because of the presence of polar bears, escorted tours are recommended.

As autumn turns bearberry and lichen into a carpet of color, opposite, the changing seasons also spur caribou, above, to move from the exposed tundra to the shelter of the forest. The animals follow routes so well-worn by millennia of seasonal movement over the coastal plains, inset opposite, that their trails can be clearly seen from the air.

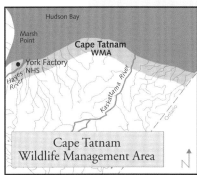

Cape Tatnam Wildlife Management Area

Designation of Lands in Manitoba

Crown Land: Crown land refers to public lands in Manitoba that are administered by the federal or provincial governments.

Ecological Reserve: Provincial Crown lands set aside for ecosystem and biodiversity preservation, research, education and nature study. Resource use, trapping, hunting and the collection of plants and animals are prohibited.

National Park: A Canada-wide system of representative natural areas of national significance. By law, they are protected for public understanding, appreciation and enjoyment, while being maintained in an unimpaired state for future generations.

National Historic Site: One of a Canada-wide system of nationally significant heritage places, presented and preserved for future generations.

Provincial Park: Provincial Crown lands set aside to conserve ecosystems and maintain biodiversity; preserve unique and representative natural, cultural and heritage resources, and provide outdoor recreational and educational opportunities in a natural setting.

Provincial Park Reserve: Provincial Crown lands set aside for potential designation in future as a provincial park.

Refuge: There are several types of refuges identified in the province's Wildlife Act. The most restrictive is the Wildlife Refuge which protects all species from hunting and trapping. Game Bird Refuges protect only game birds unless other restrictions apply to the same area. Other refuges are species specific, protecting geese or only Canada geese in some cases.

Special Conservation Area: An area established to protect rare or endangered species, primarily by restricting public access during critical times of the year.

Wildlife Management Area: Wildlife Management Areas (WMAs) are provincial Crown lands set aside for the better management and conservation of wildlife. Resource use, hunting and trapping are permitted in most WMAs.

Highway Symbols

- **1** Trans-Canada Highway
- **16** Yellowhead Highway
- **7** Provincial Trunk Highway (PTH)
- **220** Provincial Road (PR) – paved
- **367** Provincial Road (PR) – unpaved
- Other Roads

Conversions (Metric to Imperial)

When you know	Multiply by	To find
metres	3.28	feet
kilometres	0.62	miles
hectares	2.47	acres
kilograms	2.20	pounds
Celsius temperature	9/5 & add 32	Fahrenheit temperature

Suggested Reading

The Birds of Canada, by W. Earl Godfrey. Ottawa: National Museums of Natural Sciences, National Museums of Canada, 1986.

A Birder's Guide to Churchill, by Bonnie Chartier. Colorado Springs: American Birding Association, 1994.

Birder's Guide to Southwestern Manitoba, by Calvin W. Cuthbert et al. In cooperation with the Brandon Natural History Society, 1990.

Birder's Guide to Southeastern Manitoba, by Norman J. Cleveland et al. Winnipeg: Manitoba Naturalists Society, 1988.

A Field Guide to the Birds of Eastern and Central North America, by Roger Tory Peterson. Boston: Houghton Mifflin, 1980.

Field Guide to the Birds of North America, S.L. Scott, ed. Washington, D.C.: National Geographic Society, 1987.

Birds of North America: A Guide to Field Identification, by Chandler S. Robbins et al. New York: Golden Press, 1966.

Watching Birds, by Roger F. Pasquier. Boston: Houghton Mifflin, 1977.

The Butterflies of Manitoba, by Paul Klassen et al. Winnipeg: Manitoba Museum of Man and Nature, 1989.

The Mammals of Canada, by A.W.F. Banfield. Toronto: University of Toronto Press, 1974.

The Amphibians and Reptiles of Manitoba, by William Preston. Winnipeg: Manitoba Museum of Man and Nature, 1982.

Wildflowers Across the Prairies, by Fenton R. Vance et al. Vancouver: Douglas & McIntyre, 1992.

Newcomb's Wildflower Guide, by Lawrence Newcomb. Boston: Little, Brown, 1977.

Wildflowers of Churchill and the Hudson Bay Region, by Karen L. Johnson. Winnipeg: Manitoba Museum of Man and Nature, 1987.

Field Guide to the Native Trees of Manitoba, by Edward T. Oswald and Frank H. Nokes. Winnipeg: Canadian Forest Service and Manitoba Natural Resources, 1993.

Plants of the Western Boreal Forest and Aspen Parkland, by Derek Johnson et al. Edmonton: Lone Pine, 1995.

Wilderness Rivers of Manitoba, by Hap Wilson and Stephanie Aykroyd. Merrickville: Canadian Recreational Canoeing Association, 1998.

Canoeing Manitoba Rivers, Volume One, by John Buchanan. Calgary: Rocky Mountain Books, 1997.

The Geography of Manitoba: Its Land, Its People, by John Welsted et al. (eds). Winnipeg: University of Manitoba Press, 1996.

Natural Heritage of Manitoba: Legacy of the Ice Age, J.T. Teller, ed. Winnipeg: Manitoba Museum of Man and Nature, 1984.

Manitoba Travel Guide, by David Hatch et al. Winnipeg: Manipeg Publishing, 1995.

Manitoba: A Colour Guide Book, by Marilyn Morton. Halifax: Formac, 1995.

Manitoba Outdoor Adventure Guide: Cycling, by Ruth Marr. Saskatoon: Fifth House, 1989.

Manitoba Walking and Hiking Guide, by Ruth Marr. Saskatoon: Fifth House, 1990.

Some Sought-after Species & Where to Find Them

Mammals

Bear, black: Riding Mountain, Whiteshell, Duck Mountain, Grass River, Long Point

Bear, polar: Wapusk & Cape Churchill, Cape Tatnam, Seal River, Churchill River

Bison, plains: Riding Mountain

Bison, wood: Waterhen

Caribou: Northern Caribou Country, Wapusk & Cape Churchill, Cape Tatnam, Seal River, Grass River, Atikaki

Coyote: Spruce Woods, Pembina Valley, Riding Mountain, West Lake Sites

Deer, white-tailed: Fort Whyte Centre, Birds Hill, Assiniboine Park & Forest, Pembina Valley, Poverty Plains, Beaudry, Buffalo Point, Spruce Woods, Turtle Mountain, Asessippi, Brandon Hills, Souris River Bend

Elk: Riding Mountain, Duck Mountain, Spruce Woods, Mantagao Lake

Moose: Riding Mountain, Turtle Mountain, Duck Mountain

Otter, river: Whiteshell, Nopiming, Atikaki

Whale, beluga: Churchill River, Seal River

Wolf: Riding Mountain, Hecla/Grindstone, Northern Caribou Country, Wapusk & Cape Churchill, Cape Tatnam, Seal River, Grass River, Atikaki

Birds

Blackbird, yellow-headed: Oak Hammock Marsh, Hecla/Grindstone, Delta Marsh, Crescent Lake, Minnedosa Potholes & Proven Lake, Oak Lake-Plum Lakes, Grand Beach, Whitewater Lake, Alexander-Griswold Marsh, Pinkerton Lakes, Holmfield.

Bunting, indigo: Birds Hill, La Barrière Park, Beaudry, Riding Mountain, Brandon Hills, Spruce Woods, Turtle Mountain

Eider, common: Akudlik Marsh, Churchill River

Godwit, Hudsonian: Wapusk & Cape Churchill, Akudlik Marsh (migrant at Oak Hammock Marsh and Delta Marsh)

Goose, Ross's: Cape Tatnam, Seal River, Delta Marsh, Whitewater Lake, West Lake Sites, Oak Hammock Marsh

Grebe, Clark's: Delta Marsh, Pembina Valley (Pelican Lake), Oak Lake-Plum Lakes

Grebe, western: Hecla/Grindstone, Delta Marsh, Grand Beach, Pembina Valley (Pelican Lake), Oak Lake-Plum Lakes

Grouse, sharp-tailed: Oak Hammock Marsh, West Lake Sites, Narcisse, Poverty Plains, Tall Grass Prairie Preserve, Souris River Bend, Spruce Woods, Shoal Lakes

Grouse, spruce: Riding Mountain, Whiteshell, Hecla/Grindstone, Spur Woods, Porcupine Provincial Forest, Duck Mountain, Grass River

Dennis Fast

Gull, Ross's: Akudlik Marsh, Churchill River

Hawk, ferruginous: Poverty Plains

Hawk, Swainson's: Poverty Plains, Brandon Hills, Oak Hammock Marsh

Longspur, chestnut-collared: Poverty Plains, Spruce Woods

Longspur, Smith's: Wapusk & Cape Churchill; (migrant at Oak Hammock Marsh and Oak Lake-Plum Lakes)

Loon, Pacific: Akudlik Marsh, Churchill River, Wapusk & Cape Churchill

Loon, red-throated: Churchill River, Seal River

Oldsquaw: Churchill River, Wapusk & Cape Churchill (migrant at Pinawa and Seven Sisters Falls)

Owl, great gray: Whiteshell, Spur Woods, Buffalo Point, Hecla/Grindstone, Riding Mountain, Spruce Woods, Pinawa, Nopiming

Pelican, American white: Hecla/Grindstone, Oak Hammock Marsh, Delta Marsh, Grand Beach, Pembina Valley, West Lake Sites, Seven Sisters Falls, Netley-Libau Marsh, Shoal Lakes, Dog Lake & Lake Manitoba Narrows, Portia Marsh & Bluff Creek, Long Island Bay & Lake Winnipegosis Islands, Overflowing River

Pipit, Sprague's: Poverty Plains, Shoal Lakes, Oak Lake-Plum Lakes, Spruce Woods

Plover, piping: Grand Beach, Shoal Lakes, Long Point, Delta Marsh

Ptarmigan, willow: Wapusk & Cape Churchill, Cape Tatnam, York Factory

Rail, yellow: Oak Hammock Marsh, Tall Grass Prairie Preserve, Akudlik Marsh

Shrike, northern: Wapusk & Cape Churchill, Cape Tatnam, Seal River, Northern Caribou Country (also winters in many southern Manitoba sites, ie. Birds Hill, Buffalo Point)

Sparrow, Baird's: Poverty Plains, Oak Lake-Plum Lakes

Sparrow, Le Conte's: Tall Grass Prairie Preserve, Oak Hammock Marsh, Delta Marsh, Pembina Valley, Minnedosa Potholes

Sparrow, Nelson's sharp-tailed: Delta Marsh, Minnedosa Potholes, Oak Lake-Plum Lakes, Wapusk & Cape Churchill, Oak Hammock Marsh, Pembina Valley, Tall Grass Prairie Preserve

Tern, Arctic: Churchill River, Wapusk & Cape Churchill, Cape Tatnam, Seal River

Tern, Caspian: Hecla/Grindstone, Delta Marsh, Seven Sisters Falls, Long Island Bay & Lake Winnipegosis Islands.

Vireo, yellow-throated: Assiniboine River Corridor-Brandon, Turtle Mountain, Oak Lake-Plum Lakes, Asessippi, Souris River Bend, Beaudry, La Barrière, Assiniboine Park

Warbler, Connecticut: Riding Mountain, Whiteshell, Narcisse (migrant at Delta Marsh)

Warbler, golden-winged: Riding Mountain, Duck Mountain

Whimbrel: Wapusk & Cape Churchill

Woodcock, American: Birds Hill, Netley-Libau, Duck Mountain, Buffalo Point, Spur Woods

Woodpecker, black-backed: Riding Mountain, Whiteshell, Nopiming, Pinawa, Wapusk & Cape Churchill

Woodpecker, three-toed: Wapusk & Cape Churchill, Riding Mountain, Hecla/Grindstone, Whiteshell, Nopiming, Pisew Falls

Wren, sedge: Delta Marsh, Minnedosa Potholes, Oak Lake-Plum Lakes, Tall Grass Prairie Preserve, Oak Hammock Marsh, Pembina Valley, West Lake Sites, Pinkerton Lakes

Checklists

The Manitoba Naturalists Society has published checklists to the birds, butterflies and mammals of Manitoba.

Manitoba's Top Wildlife Viewing Sites

Akudlik Marsh 242
Alexander-Griswold Marsh 165
Asessippi Provincial Park 195
Assiniboine Park & Forest 19
Assiniboine River Corridor-Brandon 150
Athapapuskow Lake 220
Atikaki Provincial Wilderness Park 87
Beaudry Provincial Park 28
Big Valley 180
Birds Hill Provincial Park 36
Brandon Hills WMA 152
Buffalo Point, Birch Point
 & Moose Lake 58
Cape Tatnam WMA 250
Churchill River Estuary 238
Crescent Lake 128
Delta Marsh 110
Dog Lake WMA
 & Lake Manitoba Narrows 105
Duck Mountain Provincial Park 201
Fort Whyte Centre 22
Frank Skinner Arboretum 198
Grand Beach Provincial Park 49
Grass River Provincial Park 223
Hecla/Grindstone Provincial Park 99
Holmfield WMA 149
La Barrière Park 34
Living Prairie Museum 26
Long Point & Katimik Lake 108
Long Island Bay
 & Lake Winnipegosis Islands 117
Mantagao Lake WMA 106
Minnedosa Potholes
 & Proven Lake WMA 178
Narcisse WMA 96
Netley-Libau Marsh 94
Nopiming Provincial Park 77

Northern Caribou Country 232
Oak Hammock Marsh WMA 30
Oak Lake-Plum Lakes 166
Overflowing River 213
Pembina Valley WMA 147
Pinawa 74
Pinkerton Lakes 131
Pisew Falls Provincial Park 227
Porcupine Provincial Forest 212
Portia Marsh & Bluff Creek 116
Poverty Plains 168
Riding Mountain National Park 183
Seal River 229
Seven Sisters Falls 72
Shoal Lakes 104
Souris River Bend WMA 160
Spruce Woods Provincial Park 132
Spur Woods WMA 56
Tall Grass Prairie Preserve 125
Tom Lamb & Saskeram WMAs 214
Turtle Mountain Provincial Park 154
Vermillion Park 192
Wapusk National Park
 & Cape Churchill WMA 245
Waterhen Wood Bison 118
West Lake Sites 114
Whiteshell Provincial Park 61
Whitewater Lake 163
York Factory,
 Nelson & Hayes Rivers 248